SCI PUBLICATION 110

Slim Floor Design and Construction

Entwurf und Bau von Flachdecken

Dimensionnement et construction en planchers minces

Progetto e realizzazione di solai di limitato spessore

Cálculo y construcción de forjados esbeltos

D L MULLETT IEng MIMechIE

ISBN 1 870004 69 8

British Library Cataloguing in Publication Data
A catalogue record for this book is available from the British Library

© The Steel Construction Institute 1992

The Steel Construction Institute Silwood Park, Ascot Berkshire SL5 7QN Telephone: 0344 23345 Fax: 0344 22944	Offices also at: Unit 820, Birchwood Boulevard Birchwood, Warrington Cheshire WA3 7QZ	B-3040 Huldenberg 52 De Limburg Stirumlaan Belgium

FOREWORD

The only region of Europe other than the UK that has shown a marked increase in the use of structural steel in recent years is Scandinavia. One factor which can be attributed to this success of steel construction in Scandinavia is the so called "hat beam" used in slim floor construction. Such an innovation has brought out many alternative ideas on how to form the "hat beam". It was for these reasons that British Steel plc decided to send a team of structural engineers to Sweden to investigate the merits of this system. The findings of that investigation plus further research into slim floor construction have been used to produce this publication. This has led to the development of a slim floor system for use in the UK together with an appropriate design procedure consistent with BS 5950: Parts 1: 1990 and Part 3: Section 3.1, 1990.

The research leading to this publication was funded by British Steel General Steels. During this research British Steel and SCI developed the design of a slim floor beam, "SLIMFLOR", which is referred to extensively in this publication. An application for SLIMFLOR as a registered trademark has been made.

The author of the publication was Mr D L Mullett (SCI), and Section 5 was prepared by Mr G M Newman (SCI). The work was initiated by Mr P J Wright and Miss M MacDermott of British Steel General Steels.

CONTENTS

	Page
SUMMARY	iv
NOTATION	vii

			Page
1.	**INTRODUCTION**		1
	1.1	Background information	1
	1.2	Definitions	1
	1.3	Benefits of slim floor construction	1
2.	**ALTERNATIVE ARRANGEMENTS OF THE "HAT BEAM"**		3
	2.1	Internal beam	3
	2.2	"The SLIMFLOR" beam	4
3.	**CONSTRUCTION TECHNIQUES**		9
	3.1	Steel frames	9
	3.2	Floor construction	11
4.	**BASIS OF DESIGN**		17
	4.1	Non-Composite beam	18
	4.2	Semi-Composite beam	30
	4.3	Composite construction	32
	4.4	Edge beams	41
	4.5	Connections	41
	4.6	Robustness	42
5.	**FIRE RESISTANCE**		46
	5.1	Fire tests	45
	5.2	Required strength in fire	49
	5.3	Composite beams	49
	5.4	Design program	50
	5.5	Recommendations	50
6.	**DESIGN CHARTS FOR INITIAL SIZING**		52
REFERENCES			59
APPENDIX A: Worked examples			61
APPENDIX B: Formulae for elastic section properties and plastic moment capacity			123
APPENDIX C: User notes for the SLIMFLOR program			147

SUMMARY

Slim floor design and construction

The main aim of this publication is to present a method of design for slim floor construction comprising steel beams and concrete slabs located within the depth of the beams. The slabs are supported on a plate attached to the bottom flange of the beam. The design procedures are in accordance with BS 5950: Parts 1 (1990) Part 3: Section 3.1 (1990) for unpropped simply supported beams with uniformly distributed loading. Plastic analysis has been adopted for design of the cross-section at ultimate loads and elastic analysis for the serviceability condition. Design charts and explanation of their use for initial sizing has been included.

A computer program complementary to this publication is also available for purchase separately (see the back cover for ordering information). It is interactive and suitable for use on any IBM compatible PC. The program covers:

- Structural design for vertical loads
- Fire engineering.

The structural design part of this program performs the majority of checks presented in this design guide.

The fire engineering part of this program simulates the thermal response found in full scale tests. This information defines whether additional fire protection is required. The program will enable the designer to obtain section sizes rapidly with the facility of printing a formal set of design calculations, if required.

Three fully worked examples have been included which illustrate the design methods for non-composite, semi composite and composite action.

Entwurf und Bau von Flachdecken

Zusammenfassung

Das Hauptziel dieser Veröffentlichung besteht darin, eine Methode für den Entwurf von Flachdecken vorzustellen, die aus Stahlträgern und innerhalb der Trägerhöhe eingebrachten Betondecken bestehen. Die Betondecken leigen auf einer, an dem Unterflansch des Trägers befestigten, Stahlplatte. Der Entwurf stimmt überein mit BS 5950, Teil 1 (1990) und Teil 3 (1990), für Einfeldträger ohne zwischenstützen und Gleichstreckenlast. Plastische Berechnung bei der Querschnittsbemessung unter Grenzlasten und elastische Berechnung beim Nachweis der Gebrauchstauglichkeit wurden zugrundegelegt. Bemessungstafeln nebst Anleitung sind beigefügt.

Ergänzend zu dieser Veröffentlichung kann ein Computer-Programm käuflich erworben weden (siehe hintere Umschlagseite). Es ist interaktiv und auf jedem IBM-kompatiblen PC lauffähig. Das programm behandelt die:

- *Bemessung für Vertikallasten*
- *Bemessung im Brandfall.*

Die Bemessung im Brandfall simuliert die, anhand von Brandversuchen, ermittelten Temperaturen. Diese Information legt fest, ob zusätzliche Brandschutzmaßnahmen nötig sind.

Das Programm erlaubt dem Anwender eine schnelle Querschnittsermittlung, mit der Möglichkeit des Ausdrucks der Berechnungen.

Drei vollständige Berechnungsbeispiele sind enthalten; sie zeigen die Berechnungsmethoden für Querschnitte ohne Verbund, mit Teilverbund und vollem Verbund auf.

Dimensionnement et construction en planchers minces

Résumé

L'objectif de cette publication est de présenter une méthode de dimensionnement des planchers minces composés de poutres en acier et de dalles en béton situées dans la hauteur du profilé de poutre. Les dalles sont supportées par une plaque attachée à la semelle inférieure de la poutre. Les procédures de dimensionnement sont basées sur la norme BS 5950 Partie 1 (1990) et Partie 3 (1990) pour les poutres simplement appuyées, non étayées et chargées uniformément. L'analyse plastique a été adoptée pour le dimensionnement de la section aux états ultimes tandis que sous conditions de service, on utilise une analyse élastique. Des diagrammes de dimensionnement, ainsi que les explications nécessaires à leur utilisation, permettent un dimensionnement préliminaire (dimensions initiales).

Un programme informatique qui accompagne cette publication peut être acheté séparément (voir bon de commande à l'intérieur de la couverture). Il est interactif et utilisable sur tout ordinateur compatible IMB PC. Le programme contient:

- *Dimensionnement structural sous charges verticales*
- *Incendie*

Le dimensionnement structural réalise la majorité des vérifications présentées dans ce guide.

La partie consacrée à l'incendie simule les températures obtenues lors d'essais en vraie grandeur. Les informations obtenues permettent de décider si une protection additionnelle à l'incendie est nécessaire. Le programme permet au projeteur d'obtenir rapidement les sections nécessaires tout en lui fournissant, si nécessaire, une liste des calculs.

Trois exemples sont inclus, qui illustrent la méthode en tenant compte d'une collaboration nulle, partielle ou totale entre l'acier et le béton.

Progetto e realizzazione di solai di limitato spessore

Sommario

Lo scopo principale di questa pubblicazione e' la presentazione di un metodo di progetto per solai di limitato spessore costituiti da travi in acciaio al cui interno sono posizionate solette in calcestruzzo. Le solette sono appoggiate ad un piatto collegato all'ala inferiore della trave. I criteri progettuali riportati sono in accordo con la normativa BS 5950: Parte 1 (1990) e Parte 3 (1990) relativi alle travi in semplice appoggio non puntellate e soggette a carico uniformemente distribuito. Per il progetto delle sezioni trasversali allo stato limite ultimo sono stati utilizzati i criteri dell'analisi plastica mentre l'analisi elastica e' stata utilizzata per le condizioni di servizio. Sono inclusi abachi progettuali per il dimensionamento iniziale con le relative istruzioni di utilizzo.

Il programma di calcolo di supporto a questa pubblicazione (vedere la copertina posteriore per informazione relative al suo ordine) e' anche disponibile separatamente. Il programma, interattivo ed utilizzabile su ogni Personal Computer IBM compabile, ha i seguenti campi di applicabibilita':

- *Progettazione strutturale per i carichi verticali*
- *Comportamento al fuoco*

La prima parte permette la maggior parte delle verifiche presentate in questa guida progettuale.

Nella parte relativa al comportamento al fuoco sono simulate le temperature raggiunte in prove sperimentali su modelli a grandezza reale. In tal modo e' possibile capire se e' richiesta un ulteriore sistema di protezione al fuoco. Il programma consente al progettista di ottenere rapidamente le dimensioni delle sezioni con la possibilita' di stampare, se richiesto, i dati principali relativi alle varie fasi di calcolo.

Vengono anche inclusi tre completi casi progettuali che illustrano i metodi di progetto per i differenti tipi di interazione (composta, semi-composta e non composta).

Cálculo y construcción de forjados esbeltos

Resumen

El objetivo principal de esta publicación es la presentación de un método para proyectar forjados esbeltos formados por vigas de acero y placas de hormigón colocadas en el canto de las vigas o mediante apoyos en placas unidas al ala inferior de las mismas. Los métodos son congruentes con la Norma BS 5950: Partes 1 (1990) y 3 (1990) para vigas simplemente apoyadas, sin apuntalamiento y con carga uniformemente repartida. Para el cálculo de la sección se ha empleado el cáculo plástico con cargas últimas y el método elástico para las condiciones de servirin. Se han incluido tablas y diagramas, junto con explicaciones sobre su uso, para anteproyectos.

Se puede comprar por separado un programa que acompaña esta publicación (véase la sobrecubierta para hacer un pedido). Es interactivo e utilizable en cualquier ordenador personal IBM-compatible. El programa contients:

- *Cálculo estructural para cargas verticales*
- *Ingeniería antifuego*

La parte 1 realiza la mayoria de las comprobaciones recomendadas en la Guia mientras que la parte 2 simula temperaturas obtenidas en ensayos a escala natural, lo que permite definir en qué lugares se precisa protección antifuego. El programa permite al proyectista obtener rápidamente el dimensionamiento de secciones con la ventaja añadida de imprimir cálculos formales si ello es preciso.

Se han incluido tres ejemplos completos que ilustran el método de cálculo para las situaciones de acción compuesta, no compuesta o semicompuesta.

NOTATION

A	cross-sectional area of steel section (UC)
A_p	cross-sectional area of flange plate
B	breadth of steel beam
B_e	effective breadth of concrete slab
B_p	breadth of flange plate
b	$B/2$
D	beam depth
D_{pc}	depth of pre-cast unit
D_s	depth of in situ concrete slab above pc units
d	web depth as defined in BS 5950: Part 1
e	distance from centre line of beam to point load
f_{cu}	concrete cube strength
I_x	second moment of area of steel and composite sections
L_1	beam span
L_2	span of tie beam
M_c	plastic moment capacity of steel and composite section
M_s	plastic moment capacity of steel section (UC)
P_v	shear capacity of web to BS 5950: Part 1
p_y	design strength of steel section
R_c	compressive resistance of concrete slab
R_q	longitudinal shear resistance of shear connectors
R_p	tensile resistance of flange plate
R_s	tensile resistance of steel section
R_v	tensile resistance of web of depth d
R_w	tensile resistance of web including fillets
S	plastic section modulus
T	thickness of steel flange
t_w	thickness of steel web
t_p	thickness of flange plate
W	loading in kN
y	depth of elastic or plastic neutral axis
Z	elastic section modulus
α_e	ratio of elastic moduli of steel to concrete
σ	bending stresses
δ	imposed load deflection
δ_d	dead load deflection
δ_{sw}	deflection for natural frequency

1. INTRODUCTION

1.1 Background information

An innovative and economical form of steel construction resulting in decreasing the overall depth of the floors in multi-storey buildings has been developed in Scandinavia in recent years. This has been achieved by using "top hat" beams (also commonly referred to as "hat beams") and has led to the wide spread use of "slim floors" in that region. Many alternative techniques of forming the "hat beams" have been developed by various designers.

This publication is the result of an in-depth investigation into the merits of the various alternative systems, carried out by a team of structural engineers by visiting various buildings in Sweden. As a result, a slim floor system suitable for use in the UK and in other countries has been developed and is outlined within this publication. A design procedure consistent with BS 5950: Part 1 and Part 3: Section 3.1[1][2] is suggested and is applicable for beams subjected to uniformly distributed loads. Beams subject to concentrated loads are outside the scope of this publication.

1.2 Definitions

Slim floor construction is where the supporting floor beam is contained within the depth of the floor deck (Figure 1). This provides a solid flat slab appearance similar to reinforced concrete construction.

The original "hat beam" as its name implies, resembles the shape of a hat. This enables the floor slab to lie on either side of the "hat beam". A range of alternative "hat beams" has been developed (see Figure 2).

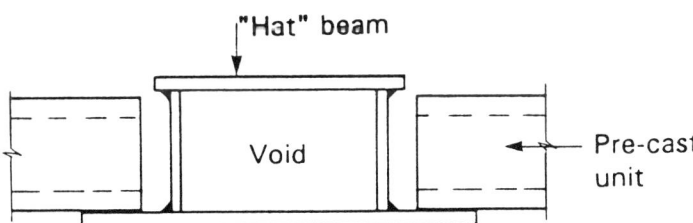

Figure 1 *Main structural floor components used in slim floor construction*

1.3 The benefits of using slim floor construction

Floors constructed using steel beams and prestressed pre-cast (pc) concrete units are usually erected with the pc units resting on the top flange of the beam. In the case of slim floors the pc units rest on the bottom flange plate as shown in Figure 1.

This simple change to the conventional method of construction produces a number of benefits:

- It offers a flat soffit to the floor.

- The overall floor construction depth can be reduced. This will reduce cladding costs.

- It improves the fire resistance of the section. The concrete that surrounds the beam partially insulates the section. This can lead to the elimination of the fire protection (see Section 5).

- The concrete that surrounds the beam produces an increase in the second moment of area of the section; this enhancement is helpful in reducing deflections

- It offers unhindered passage for services

- In the case of local element instability the concrete will improve the load carrying characteristics of the beam. For the future this could prove an asset for continuous construction.

- In certain circumstances, "dry construction" can be employed, thus saving time before the building is occupied.

2. ALTERNATIVE ARRANGEMENTS OF "HAT BEAM"

Since the inception of the "hat beam" many variations have been developed. Figure 2 illustrates some of the most recent and popular ideas to have emerged.

Figures 2(a), (b), (d) and (e) are sections frequently used in Scandinavia. The Nordic countries tend to use lighter imposed loads when compared to the UK. These light imposed loads make them less dependent on the use of in situ structural concrete to enhance the section properties of the beam. In addition, there is relatively limited production of structural universal sections in these countries and built-up plates are preferred.

The first four beams shown in Figure 2 are enclosed sections which make the task of concrete filling more difficult. The remaining beams (Figures 2e and f) which are open sections would appear to have greater flexibility for use in the UK.

2.1 Internal beams

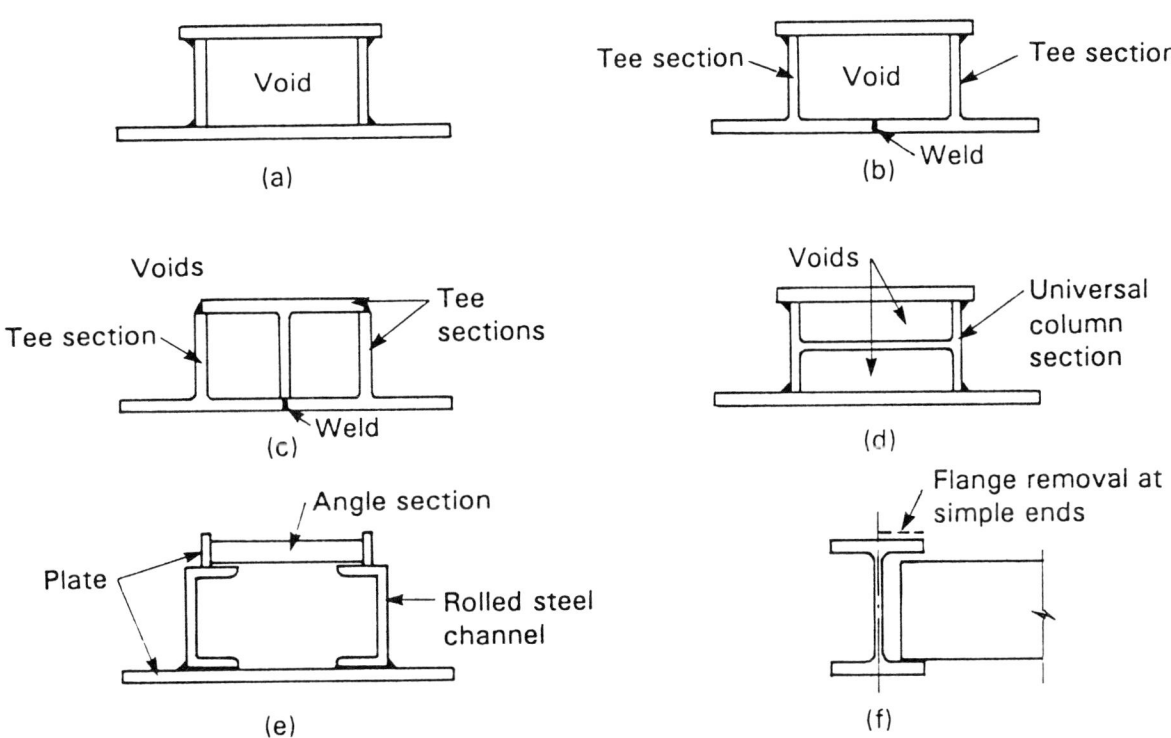

Figure 2 *Typical arrangements of internal "hat beams"*

The following is a brief description of each section:

Figure 2(a) The cross-section of the beam consists of four plates, two vertical and two horizontal giving the classic "hat" shape. This section has been used in Scandinavia notably on the domestic terminal of Arlanda Airport, Sweden.

Figure 2(b) This section is fabricated using two structural Tee sections with the flange edges touching. The stems of the Tee are vertical with a horizontal plate welded across the tops of the Tee stem.

Figure 2(c) This section is similar to Figure 2(b) the difference is that the horizontal plate is replaced by an inverted Tee section. Hence, the whole cross-section has been constructed using standard structural Tee sections.

Figure 2(d) The beam consists of a universal column (UC) section with the web horizontal and the flanges vertical. Horizontal plates are welded to the top and bottom flanges of the UC. This system is currently being fabricated in Sweden.

Figure 2(e) This form of "hat beam" is currently being marketed in the UK under the proprietary title of "ConstrucThor". The beam consists of two channels with a bottom flange plate welded to the channels. Longitudinal plates are welded to the top flange of the channel section. At intervals along the beam are cross members (Angle section) welded to the vertical plates. On plan, this produces a ladder effect which is used to transfer the horizontal shear forces into the concrete for composite action. It is also possible to use shear stud connectors in place of the vertical plates and cross members.

Figure 2(f) To enable erection of the pc units, half the width of the beam flange is reduced by notching (shown dashed). The units are then lowered through these slots and pushed along the beam. This method of construction is also explained in the SCI publication *Parallel Beam Approach - A Design Guide* [3].

2.2 The "SLIMFLOR" beam

The development of slim floor construction in the UK is still very much in its infancy. However, a few systems are available such as the ones shown in Figures 2(e) and (f). To accompany these systems, British Steel and SCI have developed a hat beam called "SLIMFLOR" (Figure 3) which utilizes a universal column (UC) section.

The UC section adopts the conventional axis (unlike the beam shown in Figure 2(d)) with a single horizontal plate welded to the bottom flange of the UC. The pc units span between beams and rest on the bottom flange plate. Figure 3 shows the basic steel components of the system.

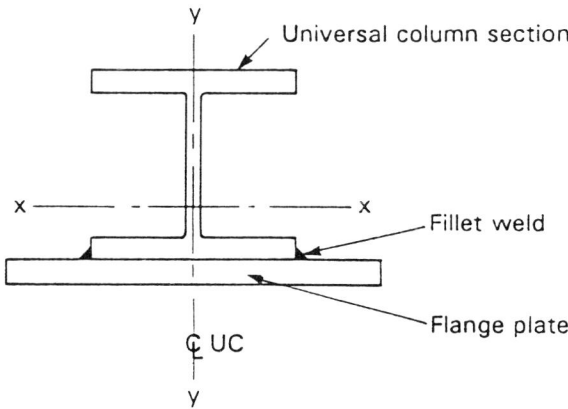

Figure 3 *The basic steel components of the "SLIMFLOR" beam*

(Note: An application for SLIMFLOR as a registered trademark has been made)

As with all slim floor systems the beam is contained within the slab depth, hence providing unhindered passage for the services. The "SLIMFLOR" beam can be easily adapted for various methods of construction, three methods are illustrated in Figure 4.

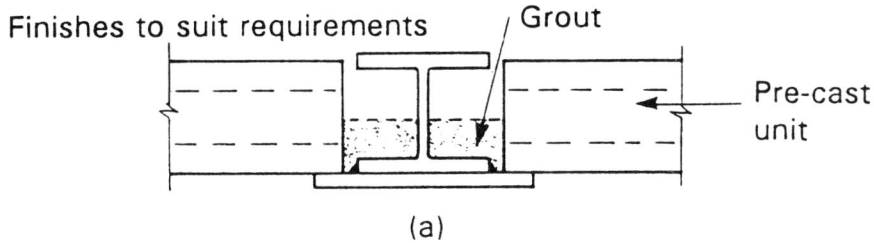

(a)

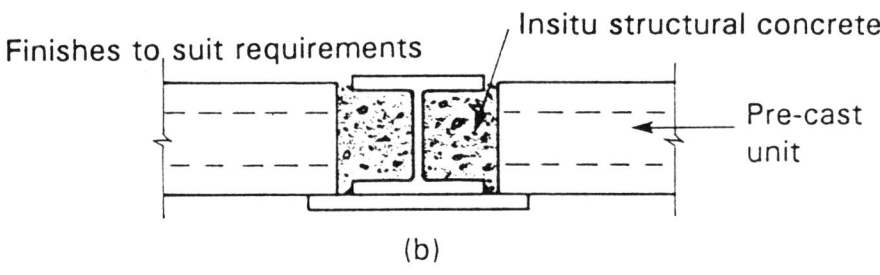

(b)

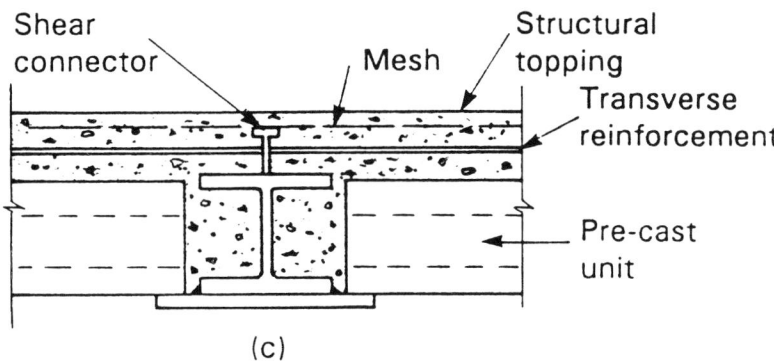

(c)

Figure 4 *Various methods of floor construction using the "SLIMFLOR" beam*

These forms of construction are fully explained in Section 3 of this publication. Also, worked examples are given in Appendix A.

The following list is a brief description of the **advantages** of the "SLIMFLOR" beam when compared to conventional forms of construction:

- It uses standard universal column sections.

- The beam is easy to fabricate with full depth end plate connections. Only two fillet welds are required to attach the longitudinal plate which can be automatically welded without turning the section.

- The system provides relatively long spans with minimum construction depths. This will have the effect of reducing cladding costs.

- No internal voids for sound or heat transfer in fire are created. This reduces the amount of fire protection.

- The system is inherently versatile to suit the requirements of a given building. This is emphasized by the three forms of construction shown in Figure 4 which range from dry to composite construction.

Note: Torsional effects in the construction stage and possibly at the edges of the building may occur under eccentric loading. Guidance has been given to eliminate or simplify these problems.

Figure 5 (a) shows a typical example of a Universal Column section with a plate welded to the bottom flange. A 254 x 254 UC section has been used with 250 mm deep pc units. The stud connectors are welded to the top flange of the UC in the fabricators works, this is highlighted in Figure 5 (b). Composite action is achieved by combining the in situ concrete with the shear connectors.

Figure 5(a) *Typical Example of the SLIMFLOR beam in the construction stage*

Figure 5(b) *Welding of steel shear connectors to the top flange of the Universal Column section*

Figure 6(a) shows the slim floor construction using the parallel beam approach; the pc units are supported on top of the bottom flange of the spine beams. A typical plan layout is shown in Figure 6(b) for this method of construction (see Reference 3 for further information).

Figure 6(a) *Slim floor construction using the parallel beam approach*

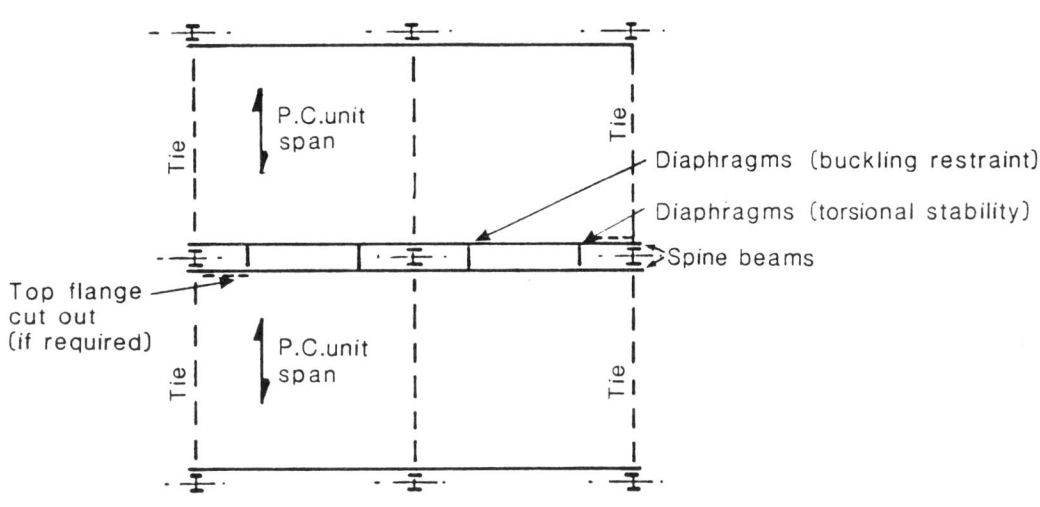

PART PLAN

Figure 6(b) *Typical plan layout for the construction shown in Figure 6(a)*

Figure 7 shows the pc units resting onto the bottom flange of the modified ConstrucThor beam. Channels are generally used to form the section (see Figure 2(e)) but in this case the channels have been replaced by vertical web plates which are continuously welded to the bottom flanges. Intermittent welds have been used to connect the web plate to the angle section. The advantage of using this form of construction is that the web plate thickness can be adjusted to suit the design loading.

Figure 7 *Modified ConstrucThor beam*

3. CONSTRUCTION TECHNIQUES

3.1 Steel frames

The construction techniques described in this section refer to the use of the SLIMFLOR system discussed in Section 2.2. Many of the principles outlined here are also applicable for other systems used to form slim floors.

The recommendations given in Table 1 are an approximate guide as to where the three basic types of construction are best used. However, comparative designs may be carried out using the software referred to in this publication, leading to a clearer understanding of the relative economy of these systems.

Table 1 *Guide to the typical uses of the "SLIMFLOR" beam*

Item \ Floor Construction	Type (1)	Type (2)	Type (3)
Description	non-composite	semi-composite	composite
Usage	Domestic	Domestic/Offices	Offices
Imposed Loads (kN/m^2)	2.5 to 3.5	3.5 to 4.5	5 to 6
Spans (m)	4 to 6	6 to 9	7 to 10
Fire Rating (minutes)*	30	60	60 to 90

* In some instances additional fire protection will be required, see Section 5.

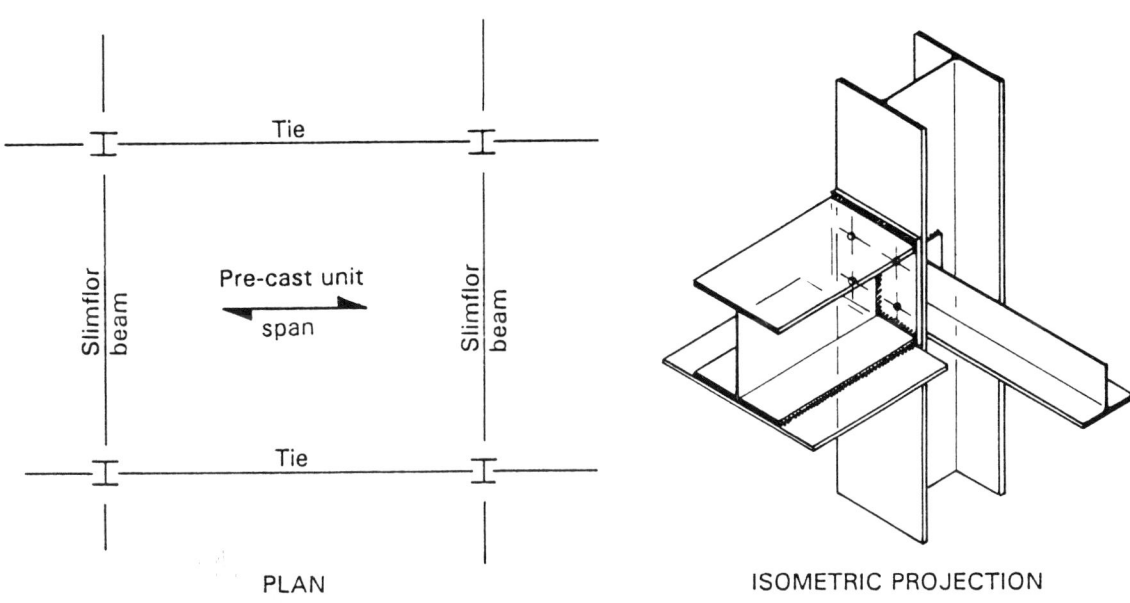

Figure 8 *Typical frame arrangement for an internal bay*

Generally, it is advantageous for the beams to span the shorter distance between the columns and the pc units to provide the larger span. By adopting this method of framing the beams become less vulnerable to deflection and vibration. Also, it enables the depth of beam to be similar to the pc unit depth (as explained later). The beams are constructed as described in Section 2 with the bottom flange plate having a thickness of approximately 15 mm. The plate thickness is usually governed by the requirements for fire resistance (see Section 5). The pc units are built so that the underside of the unit rests on the bottom flange plate. This plate extends 100 mm (minimum) either side of the UC bottom flange to provide sufficient bearing for the pc unit. The beams are then bolted (via end plate connections) to the column flange. These procedures should simplify the connections to the columns (See Figure 8).

In the UK it is necessary to provide a tie between the steel columns. This tie spans in the same direction as the pc units and is needed to satisfy the "robustness" clauses of the Building Regulations[4]. Interestingly, an internal tie is not obligatory in Scandinavia. The choice of tie member largely depends on the setting-out of the pc units. Figure 9 shows various methods of constructing the tie beam within the depth of the floor deck. The tie members are connected to the column web.

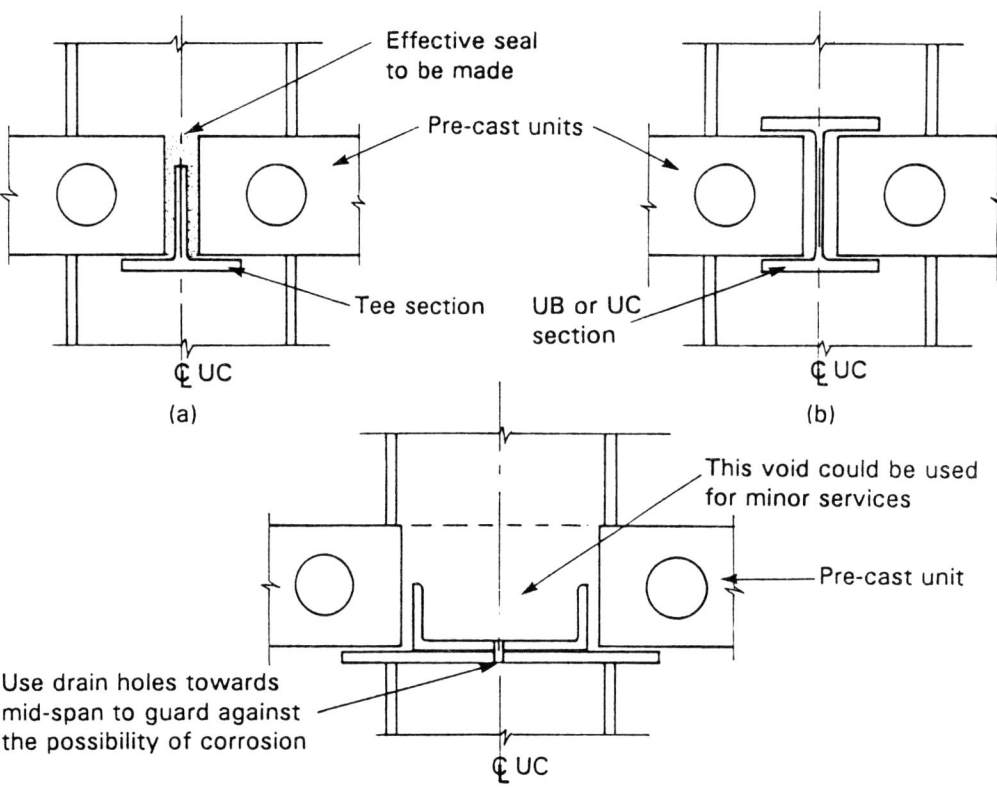

Figure 9 *Typical arrangements of tie beams*

3.2 Floor construction

Comprehensive details are given for each flooring concept (as shown in Table 1) but before any such design is finalized it is always advisable to ascertain the availability of suitable pc units from the manufacturer.

3.2.1 Internal "SLIMFLOR" beam

Type (1) Floor construction: non-composite

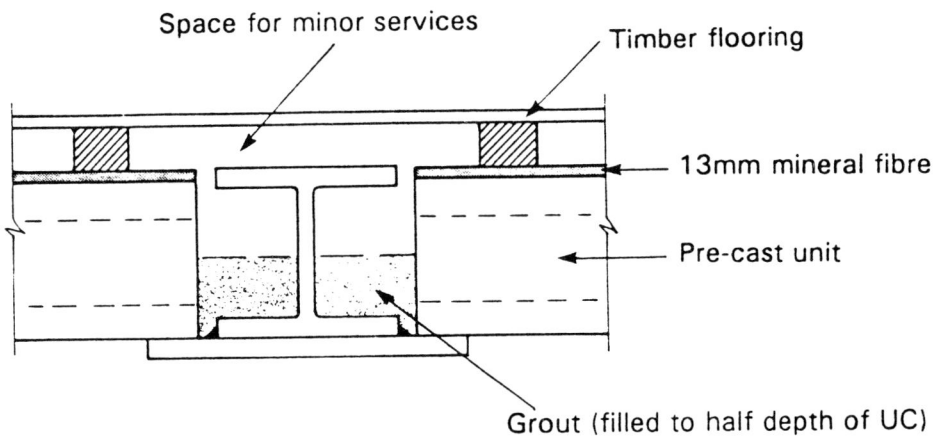

Figure 10 *Typical details for Type (1) floor construction*

A 30 to 60 minute fire rating (Section 5) for low rise domestic dwellings up to two storeys, provided each floor area does not exceed 500 m^2, would be appropriate for Type (1) floor construction. One stipulation when using this form of construction is that grout or some other suitable material must be placed around the beam. The grout should be tamped down so that an effective seal has been formed between the beam and the pc units (see Figure 10). This is to overcome the possibility, in the event of a fire, of smoke and toxic gases passing through the floor. This operation could be carried out at the same time the pc units are grouted together. The floor could be finished using traditional type timber/chipboard flooring which would eliminate the need for screeding. In addition, the timber floor could provide sufficient space for the passage of electrical services etc.

The minimum dimension between the end of the pc units and the upper flange for effective grout placement will be dependent on the equipment used and is typically 25 mm. The minimum bearing of the pc units on the steel plate is 75 mm. These dimensions dictate the width of plate that is used.

Type (2) Floor construction: semi-composite

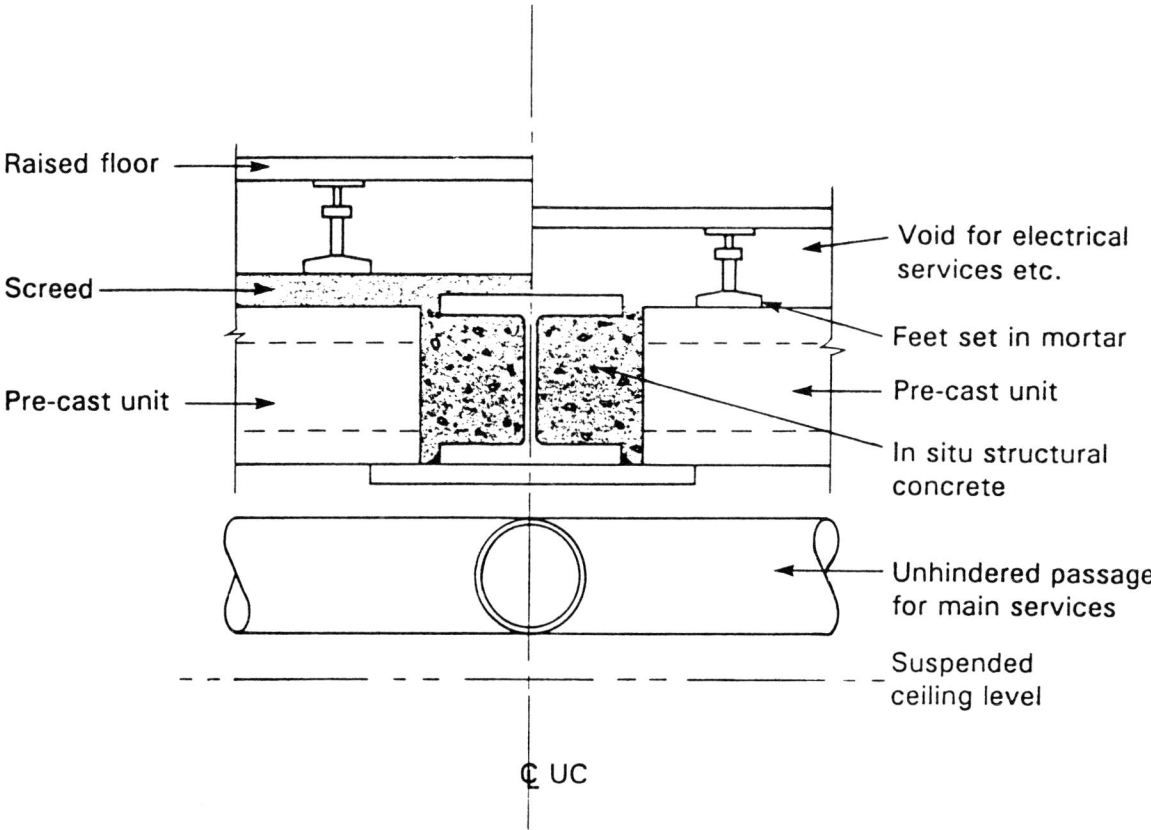

Figure 11 *Typical details for Type (2) floor construction*

This type of flooring detail would be suitable for domestic (over two storeys) or light office construction where the required fire rating (Section 5) does not exceed 60 minutes.

As previously explained the units are grouted together and in situ structural concrete (grade 30) is placed around the SLIMFLOR beam. This encasement of the beam is used for fire resistance and stiffness purposes only. In some cases a screed (with or without raised floor) is used to level the floor and also to cover reinforcement used for tying the units together. Alternatively, a raised floor may be considered where the feet that support the raised floor are set in mortar (see Figure 11). This has the attraction of dispensing with the screed and reducing construction depths. The main services have unhindered passage throughout the floor area. The ceiling which screens the services is supported using standard fixings supplied by the pc unit manufacturer.

In Type (2) and Type (3) construction the concrete encasement around the section partially insulates the steel section. Despite the high temperature of the exposed bottom plate, 60 minutes fire resistance can be readily achieved. The fire resistance can be increased by providing additional protection (by spray, board or intumescent coating) to the bottom plate.

Type (3) Floor construction: Composite

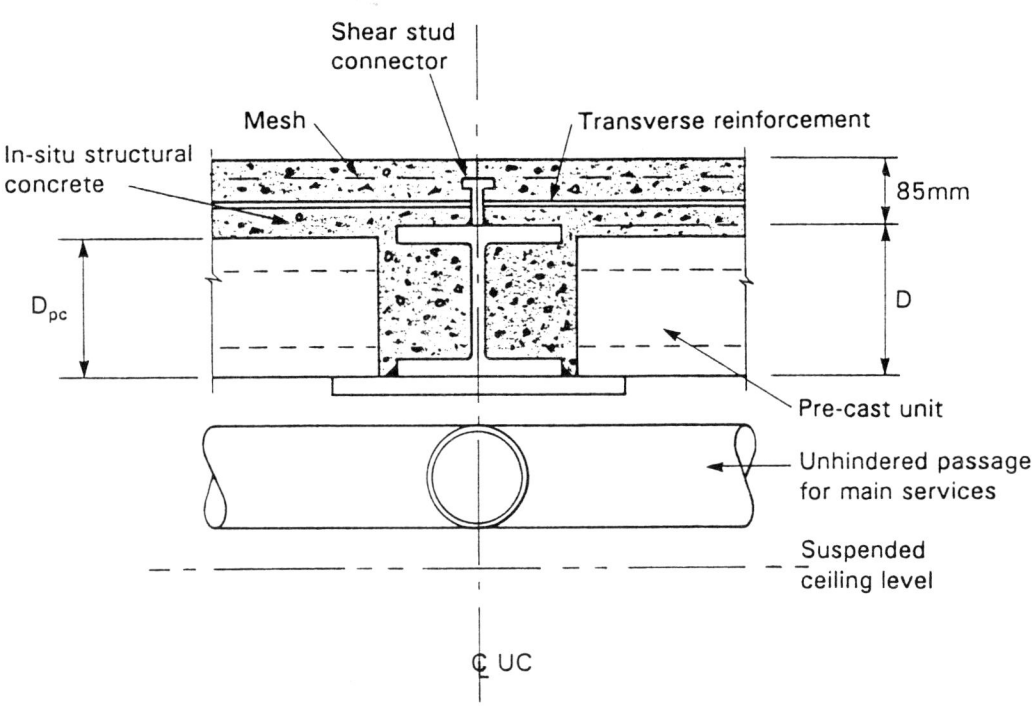

Figure 12 *Typical details for Type (3) floor construction*

In this system the steel section and pc concrete units act compositely as shown in Figure 12. Based on experience of similar structures built in Scandinavian countries and in the UK, it is envisaged that this form of construction would be most suitable where the imposed loads are high and the fire rating is 60 minutes. The construction details are similar to the ones shown in the previous example. A structural concrete topping has been added which is designed to act compositely with the steel section. For this action to be achieved, 19 mm diameter studs are shop welded to the flange. For a stud height of 70 mm (after welding) and 15 mm cover, the overall slab depth above the UC top flange is 85 mm. It is an important consideration when using this type of construction for composite action to keep the pc unit depth (D_{pc}) and the beam depth (D) similar. Otherwise, if $D_{pc} < D$ then the amount of in situ structural concrete will increase or conversely, if $D_{pc} > D$ then the overall construction depth will increase.

3.2.2 Building perimeter

Figure 13 shows five alternative methods of edge beam construction. If cladding details permit, then a downstand at the edge of the building would be the simplest form of construction (see Figure 13(e) and Section 4). These diagrams are self explanatory but a few points require expanding. Figure 13(a) shows a rolled steel channel (RSC) welded to a bottom flange plate. The pc unit bears on to the plate in a similar manner to an internal beam. Use of this form of construction eliminates torsion as the vertical loads will act approximately on the shear centre of the channel (see Section 4.4). The opposite can be said for Figures 13(b), (c) and (d) where the effects of torsion will have to be considered. If one of these three details is to be used, the enclosed section shown in Figure 13(d) would be the most favourable in terms of torsional efficiency.

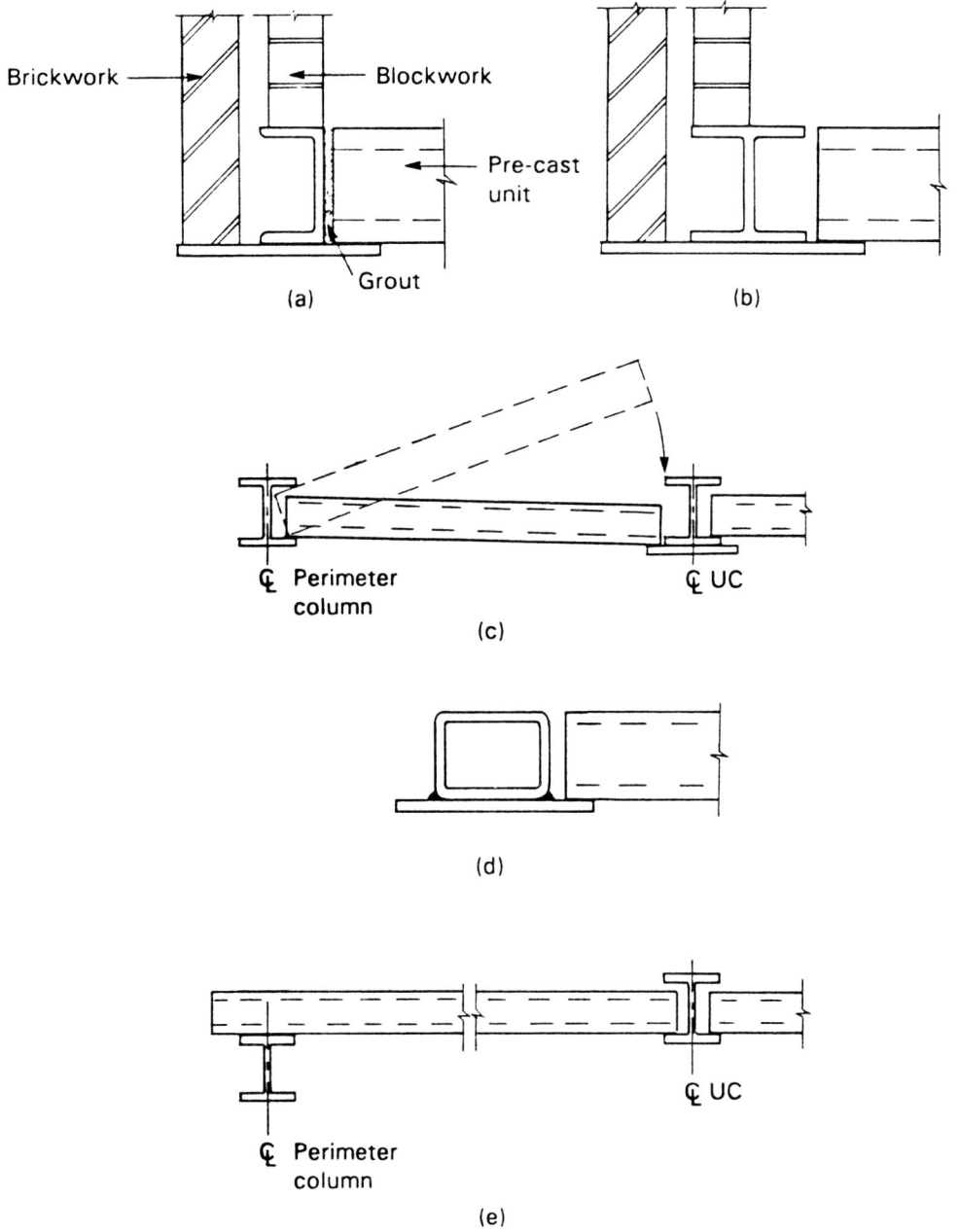

Figure 13 *Alternative arrangements of edge "hat beams"*

3.2.3 Prestressed concrete floor units

PC concrete floor units range from 100 mm to 300 mm in depth and can span up to 12 m with a maximum imposed load of 10 kN/m² (see manufacturers' literature).

The units are prestressed (to improve their load carrying capacity) and cut to length in the factory. The maximum span: depth ratio of the pc unit recommended for minimising deflections is in the range of 35 to 40. Therefore, the depth of a 7 m span slab would be typically 200 mm. The units can provide up to 2 hours fire resistance.

Figure 14(a) shows typical length tolerances and Figure 14(b) gives an indication of the unit's pre-camber. The camber will reduce by approximately 3 mm for every 1 kN/m² load.

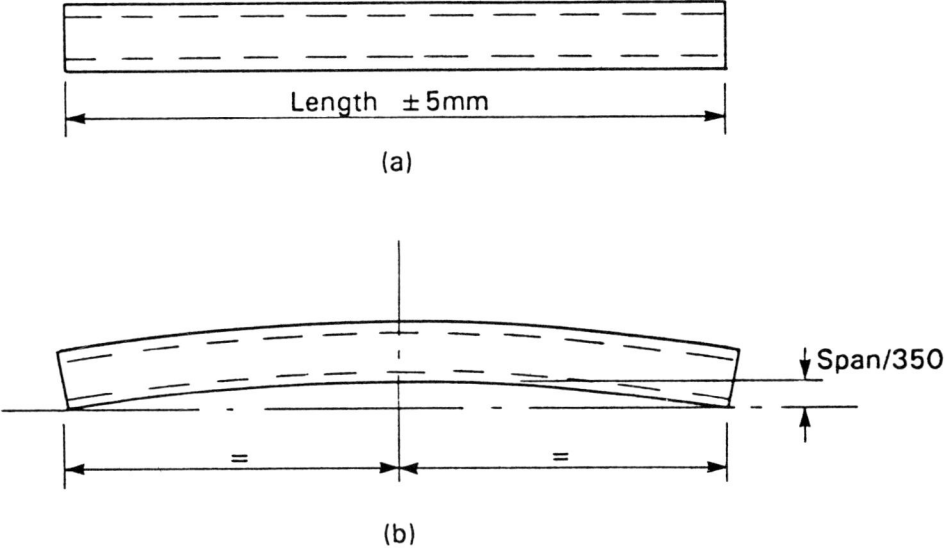

Figure 14 *Limiting dmensions for pc units*

3.2.4 Future developments

Deep steel decking

A UK Manufacturer of profiled sheeting is currently developing a steel deck which can span 6 m unpropped and has a depth of 210 mm. The deck is an ideal replacement for the pc units as it will reduce the dead weight of the floor. This has the effect of reducing steel weight and the size of foundations. Also, it lends itself for passing minor services through the slab depth between the ribs of the deck (Figure 15).

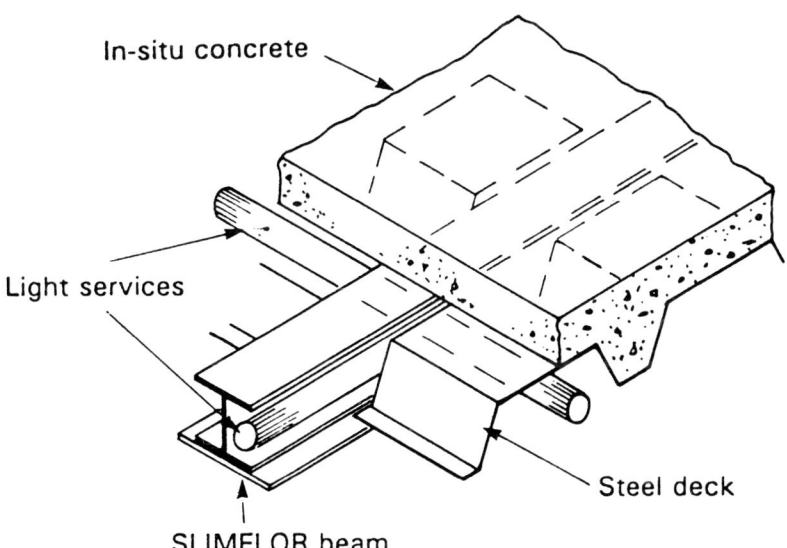

Figure 15 *Isometric projection of the deep steel deck used in composite construction*

Asymmetric rolled sections

Current research into this topic would suggest that there is a market potential for this type of section. However, the differences in flange width would need to be of the order of 150 mm to be practical for slim floor construction.

Continuous construction

Continuous construction coupled with integrated column design could produce greater economies in the use of the steel section, principally by reducing deflections.

Secondary beams

SLIMFLOR beams span between columns. It may also be possible to devise a system of secondary and primary beams where slab spans are otherwise too large to span between the SLIMFLOR beams. Primary beams are subject to point loads and will generally be deeper and heavier than the secondary SLIMFLOR beams.

Prestressed concrete floor units

The pc unit could be cast in a certain way so as to incorporate services within the unit (Figure 16(a)). Different end shapes would help the pouring of the concrete around the beam as shown in Figure 16(b).

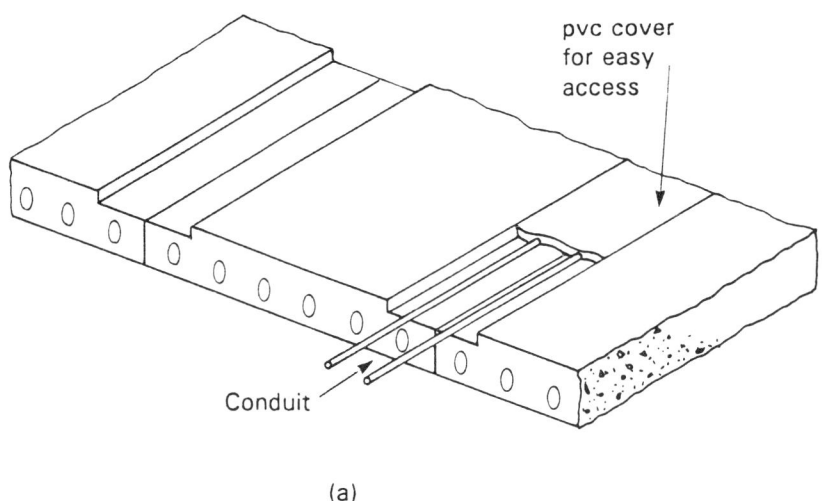

(a)

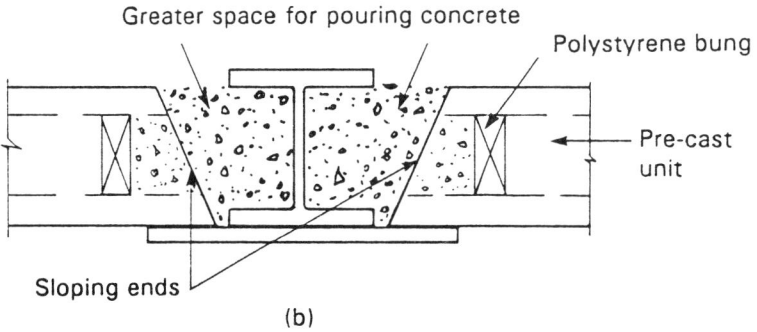

(b)

Figure 16 *Possible development of pc units*
(Note: grout will fill the holes in the pc units up to the polystyrene bung)

4. BASIS OF DESIGN

The three proposed methods of construction shown in Section 3 will be used to illustrate the design concept. The design procedures are based on the use of BS 5950: Part 1 and Part 3: Section 3.1[1][2] for unpropped simply supported beams with uniformly distributed loading.

The cross-sections will be limited to plastic (Class 1) or compact (Class 2) sections. Semi-compact (Class 3) sections would limit the design to the elastic moment capacity. This would complicate the design procedures but more importantly would prove to be an uneconomical use of the steel and semi-compact sections are therefore not recommended. In addition, BS 5950 : Part 3 : Section 3.1 (see note below Clause 2.1.3.3) clearly recommends plastic design of cross-sections irrespective of whether the global analysis is elastic or plastic.

Slim floor beams may be subject to significant torsion due to out of balance loading either during construction or in-service. Torsional effects therefore need special attention, unlike conventional construction.

The main design assumptions are:

(a) Unpropped simply supported beams are subject to uniformly distributed loading.

(b) Use only plastic or compact cross-sections.

(c) Plastic analysis of the cross-section is based on rectangular stress blocks.

(d) Moments and forces are determined using factored loads.

(e) Serviceability checks are determined using unfactored loads. To ensure that irreversible deformation (under normal service loads) does not occur in the steel, the extreme fibre stress is limited to p_y. The in situ concrete stress is likewise limited to $0.5f_{cu}$. The formulae for elastic section properties and plastic moment capacities are given in Appendix B.

(f) Deflections of beams are limited to span/360 under imposed loads and span/200 under total load. Pre-cambering should be considered when the total deflection exceeds this limit. These limitations for deflection apply to buildings of general usage. In addition, it is a requirement of BS 5950: Part 1 that due allowance should be made where deflections under serviceability loads could impair the strength or efficiency of the structure or its components or cause damage to the finishings.

For initial sizing of SLIMFLOR beams, Table 1 with the design charts presented in Section 6 should provide sufficient information. The software referred to in this publication can be used to refine the design.

4.1 Non-Composite beams

4.1.1 Design assumptions

(a) The SLIMFLOR beam (see Figure 17) is considered to be laterally restrained in the construction stage. The loading through the bottom flange constitutes a stabilizing effect.

(b) The SLIMFLOR beam is considered to be laterally restrained under imposed loading. The partially filled void around the section reduces the slenderness of the web. Tests on continuous composite beams under hogging moment (an analogous case) have shown that adequate restraint is provided if the tension flange is held in position.

(c) Out of balance loads have to be considered in the imposed load condition for Type 1 construction only. Continuity is not developed across the section and torsional effects are created. The section is therefore to be designed for combined bending from vertical loads and torsion. No simple design procedures are given for this check and the full rigorous analysis has to be carried out. (See Appendix A - worked example 1).

(d) Out of balance loads do not have to be considered in the imposed load condition for Types 2 and 3 construction.

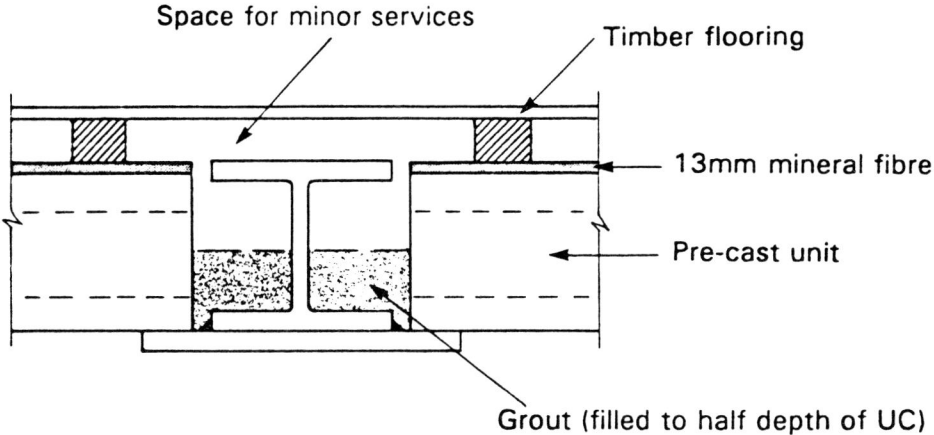

Figure 17 *Typical floor arrangement for the non-composite beam*

4.1.2 Section classification

The cross-section will be limited to plastic or compact Universal Column sections for reasons already explained.

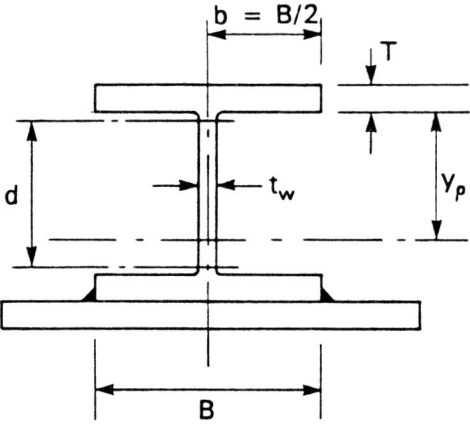

Figure 18 *Section Dimensions*

The limiting width to thickness ratios for a compact section are:

Flange $\dfrac{b}{T} \leq 8.5\epsilon$

Web (Generally) $\dfrac{d}{t_w} \leq \dfrac{98\epsilon}{\alpha}$

where: $\alpha = \dfrac{2y_p}{d}$

$\epsilon = 1.0$ for grade 43 steel
$\epsilon = 0.88$ for grade 50 steel.

If $\alpha \approx 2$ (y_p tends to d) then the section should be taken as having compression throughout. The limiting classification check then becomes:

$\dfrac{d}{t_w} \leq 39\epsilon$

All Universal Column sections of grades 43 and 50 satisfy this criterion. However, this may not apply to other shaped sections.

Table 2 shows the section classification (subject to bending) for the range of Universal Columns in grades 43 and 50 material and contains some semi-compact sections as determined by the proportions of the flanges; these are included for the information of the designer but should not be used for 'SLIMFLOR' construction.

4.1.3 Moment capacity including lateral torsional buckling effects

For the non-composite beam lateral torsional buckling (LTB) has to be considered for the construction stage unless some device is used to restrain the compression flange. Where the compression flange is considered as restrained, the full moment capacity can be developed, (see Appendix B).

The design procedures for LTB of singly symmetric sections are defined in BS 5950: Part 1[1]. At the time of writing, the section properties cannot be obtained from published tables hence, the design concept will be explained in full.

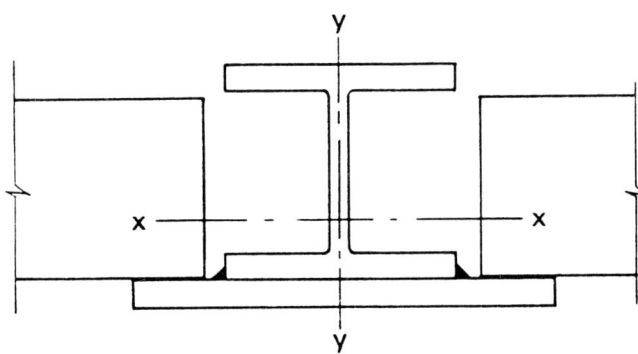

Figure 19 *Construction stage components*

Table 2 *Section classification for Universal Columns in bending for grades 43 and 50*

Serial size mm	Mass per Metre kg	Material Grade 43	Material Grade 50
356 x 406	634	P	P
	551	P	P
	467	P	P
	393	P	P
	340	P	P
	287	P	P
	235	P	P
356 x 368	202	P	P
	177	P	C
	153	C	SC
	129	SC	SC
305 x 305	283	P	P
	240	P	P
	198	P	P
	158	P	P
	137	P	P
	118	P	C
	97	SC	SC
254 x 254	167	P	P
	132	P	P
	107	P	P
	89	P	P
	73	C	SC
203 x 203	86	P	P
	71	P	P
	60	P	P
	52	P	C
	46	C	SC
152 x 152	37	P	P
	30	P	C
	23	SC	SC

Section classification:

P = Plastic
C = Compact
SC = Semi Compact

* Note: SC sections are not covered in this publication

Buckling resistance moment, M_b

$$M_b = S_x \cdot p_b$$

where:

S_x = the plastic modulus of the section about the x-x axis. This can be calculated from first principles using the stress block method. However, the simplest way to calculate S_x is to use the formulae for moment capacity (see Appendix B)

p_b = the bending strength of the section, which is related to the equivalent slenderness, λ_{LT}, the design strength of the material, p_y and member type which in this case is classed as "rolled". Once λ_{LT} has been established p_y and Table 11 (Part 1)[1] are used to evaluate p_b. Table 12 (bending strength p_b for welded sections) is not used because the welding takes place in the tension zone and the upper compression flange is part of the rolled section.

The equivalent slenderness, λ_{LT}

$$\lambda_{LT} = nuv\lambda$$

where:

n, slenderness correction factor

From Table 13 (Part 1), [1] n = 1.0, conservatively

u, buckling parameter

For flanged sections symmetrical about the minor axis

$$u = \left[\frac{4 S_x^2 \gamma}{A^2 h_s^2} \right]^{1/4}$$

where: S_x is as described above

A = cross-sectional area

$h_s \approx D - T/2$

$\gamma = (1 - I_y/I_x)$

I_y = second moment of area about the y-y axis

I_x = second moment of area about the x-x axis.

Generally, the I_x is the major axis of the section but when using this form of construction it is not always the case. With the plate being welded to the bottom flange of the UC it is possible for $I_y > I_x$. In this instance as I_y tends to I_x the γ factor becomes zero. When this happens the buckling parameter, u also becomes zero. Hence, λ_{LT} is equal to zero which nullifies the requirements for LTB and failure by LTB does not occur.

v, slenderness factor

Two methods exist for obtaining the value of v, either by using Table 14 (Part 1) [1] or alternatively the expression given for v in Appendix B of BS 5950 : Part 1. The following explains both methods:

(i) Using Table 14 (Part 1)[1]

   To use Table 14 three parameters require evaluation:

   (a) Torsional index, x
   (b) Flange ratio, N
   (c) Minor axis slenderness, λ

(a) Torsional index, x

$$x = 0.566\, h_s\, (A/J)^{1/2}$$

where: h_s and A are as previously explained.

$$J = J_{uc} + \tfrac{1}{3}(t_p^3\, B_p)$$

t_p = flange plate thickness

B_p = flange plate width

The torsion constant, J_{uc} can be obtained from published tables.

(b) Flange ratio, N

$$N = \frac{I_{cf}}{I_{cf} + I_{tf}}$$

where: I_{cf} = the second moment of area of the compression flange about the y-y axis.

$$I_{cf} = TB^3/12$$

I_{tf} = the second moment of area of the tension flange about the y-y axis.

$$I_{tf} = \frac{TB^3}{12} + \frac{t_p B_p^3}{12}$$

(c) Minor axis slenderness, λ

$$\lambda = L_E/r_y$$

where:
L_E = the effective length of the member which may be taken as $L_E = 1.0 L_1$. The loads are applied through the tension flange which represents a stabilizing condition. However, no account is taken of this beneficial effect.

r_y = the radius of gyration about the minor axis of the member.

ii) Using the formulae for v given in Appendix B of BS 5950 : Part 1 [1]

$$v = \left\{ \left[4N(1-N) + \frac{1}{20} \left(\frac{\lambda}{x} \right)^2 + \psi^2 \right]^{1/2} + \psi \right\}^{-1/2}$$

The only parameter to be determined is the monosymmetry index, ψ

$\psi = 0.8 (2N - 1)$ for $N > 0.5$
$\psi = 1.0 (2N - 1)$ for $N < 0.5$

The above information enables the designer to obtain the buckling resistance moment M_b. To illustrate these procedures a worked example has been included in Appendix A.

4.1.4 Biaxial stress effects in the flange plate

Biaxial stresses have to be considered as a direct result of the way the loads are applied to the flange plate. The plate is subject to longitudinal and transverse effects. The longitudinal stress due to overall bending of the section, σ_1, has an influence in reducing the resistance of the plate when also subject to a transverse bending stress, σ_2. This is irrespective of whether the stresses are plastic or elastic. For further information see Reference 5.

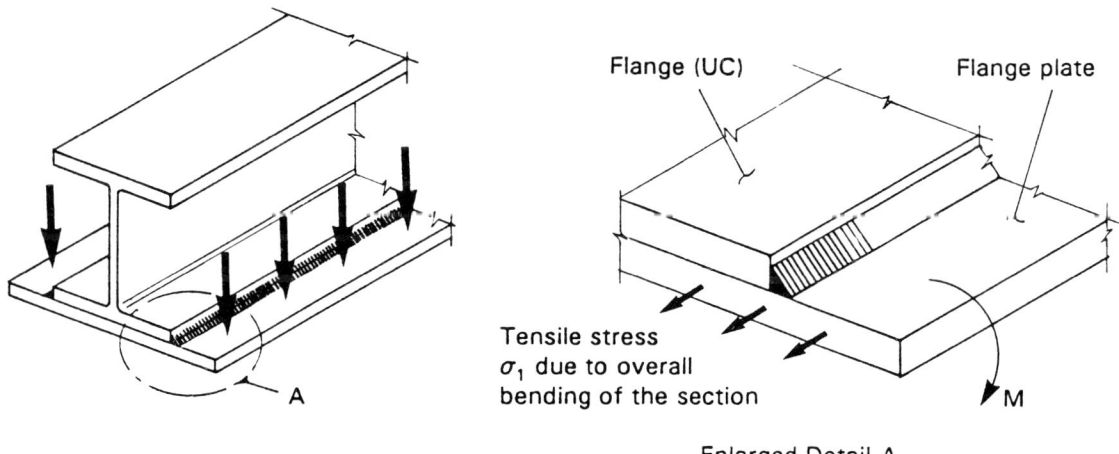

Figure 20 *Diagrams showing applied loads to the bottom flange plate*

Plastic Analysis

The effective transverse yield stress that may be resisted is reduced from p_y to σ_2 which according to Von Mises yield criterion is given by:

$$[\sigma_2^2 - \sigma_1 \sigma_2 + \sigma_1^2]^{1/2} = p_y$$

or $\quad \sigma_2 = \dfrac{\sigma_1 \pm (4p_y^2 - 3\sigma_1^2)^{1/2}}{2}$

Taking a positive sign for compression

$$\sigma_2 \text{ (compression)} = \frac{\sigma_1 + (4p_y^2 - 3\sigma_1^2)^{1/2}}{2}$$

$$\sigma_2 \text{ (tension)} = \frac{\sigma_1 - (4p_y^2 - 3\sigma_1^2)^{1/2}}{2}$$

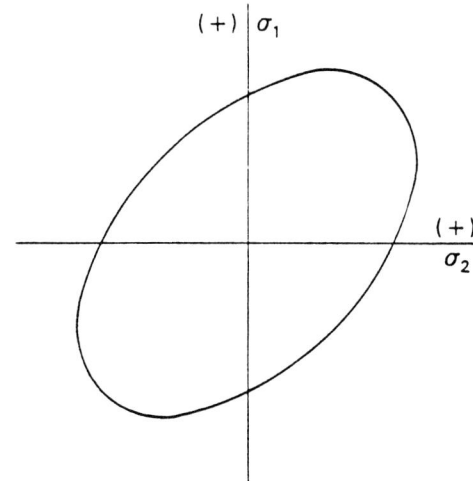

Figure 21 *Von Mises curve showing the position of σ_1 and σ_2 stresses*

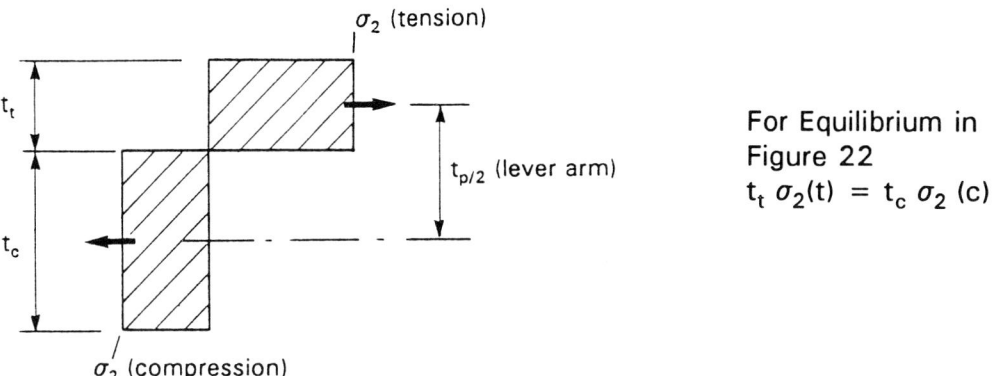

For Equilibrium in Figure 22
$t_t \, \sigma_2(t) = t_c \, \sigma_2(c)$

Figure 22 *Plastic distribution through the flange plate*

Using the above equations it can be shown that:

$$\frac{M}{M_p} = \frac{c^2 - \sigma_1^2}{2 c p_y}$$

where: $c = (4p_y^2 - 3\sigma_1^2)^{1/2}$

M = maximum transverse moment applied to the plate

M_p = moment capacity of the plate = $\dfrac{T^2 p_y}{4}$

The graph shown in Figure 23 plots (M/M_p) on the vertical axis and $(-\sigma_1/p_y)$ on the horizontal axis. This direct relationship between these two ratios gives the engineer a visual indication of the extent to which the longitudinal stress, σ_1, will influence the transverse bending stress, σ_2.

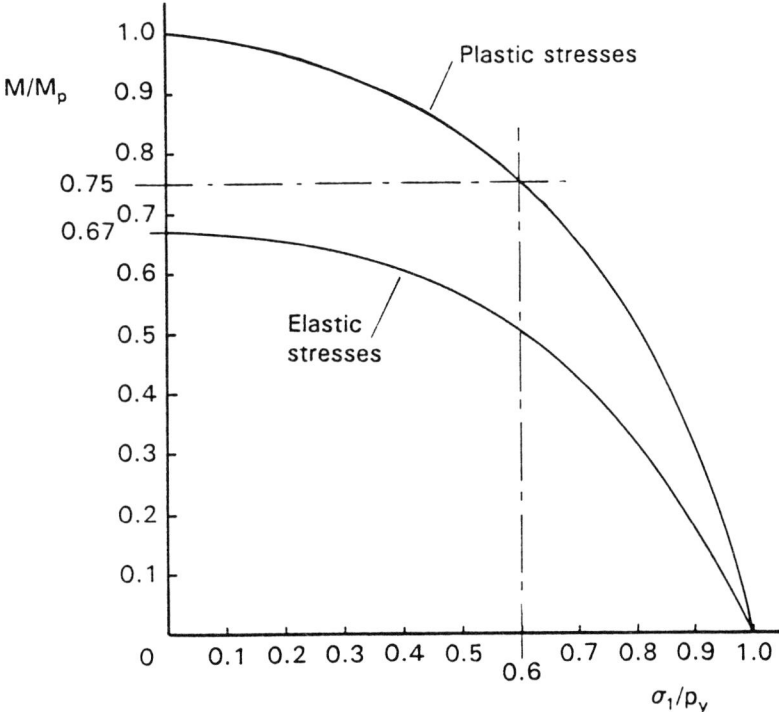

Figure 23 *Graph showing the influence of the longitudinal stress on the transverse bending stress for the plastic and elastic conditions*

Example:

Say $\sigma_1 = 0.6\, p_y$ due to overall bending, then the plate transverse moment capacity is

$\approx 0.75 M_p$ which equals $\dfrac{3}{4} \cdot \dfrac{T^2 p_y}{4} = \dfrac{3}{16} \cdot T^2 p_y$ per unit length of beam.

The above procedures have been included in the worked examples shown in Appendix A.

Elastic Stresses

The analysis in Section 4.1.3 can easily be modified to take into account the elastic stress condition. The horizontal axis plots the ratio of longitudinal stress, σ_1 to the yield stress, p_y. This ratio remains the same irrespective of the design approach. To convert the graph shown in Figure 23 account has to be taken of the difference of plate moment capacities i.e. plastic to elastic capacity.

The ratio of the plastic to elastic section modulus equals 1.5.

Therefore, the vertical axis ratio has to be limited to 0.67 $\left[\dfrac{M}{M_p}\right]$.

Using the above example for $\sigma_1 = 0.6\, p_y$ the elastic moment capacity is $\approx 0.5\, M_p$.

4.1.5 Torsional effects

In the construction stage it is not always possible to ensure that the pc units are placed in such a way as to eliminate out of balance loads. The complexities introduced by torsion are best avoided in building structures. However, in this instance this is not easily achieved unless strict erection procedures are adhered to. Out of balance loads will also have an undesirable influence on the lateral torsional buckling of the section.

The combination of these two effects has been dealt with in the SCI publication *Design of members subject to combined bending and torsion*[6]. A design example shown in Appendix A illustrates the design procedures shown in this publication. However, this analysis is rather complex, and requires simplification for general use. The following is a set of conservative design procedures which will determine the limits of use for a given grid layout.

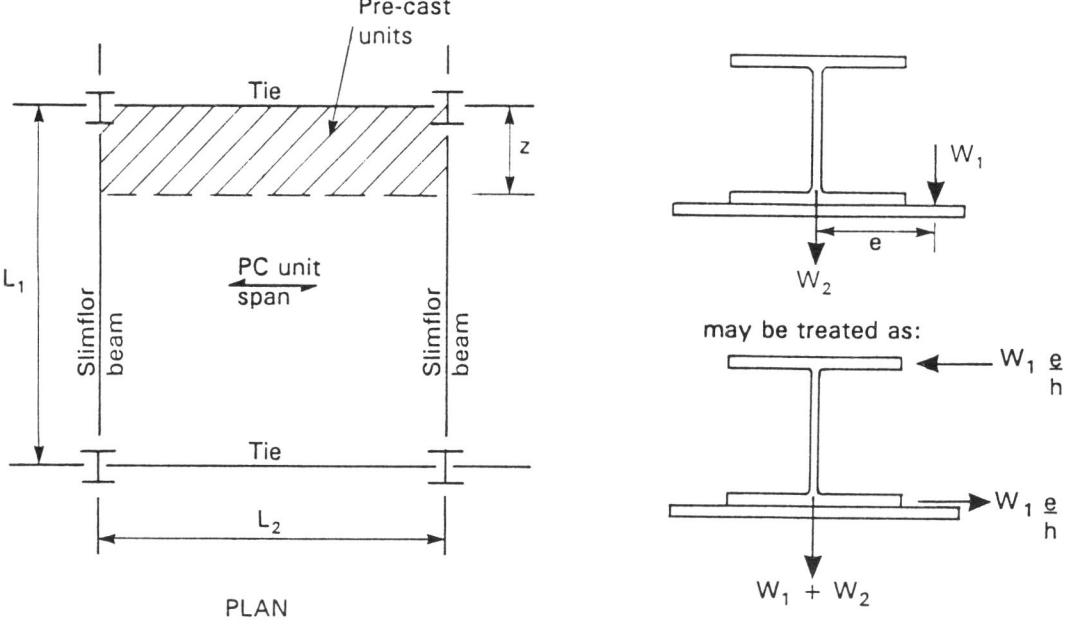

Figure 24 *Partially loaded (pc units) Internal Bay*

Figure 24 shows a plan layout for an internal bay with a typical cross-section through the floor beam. The dimension z and the enclosed hatched area is the extent of the pc units before the combined effects of lateral torsional buckling and torsion becomes onerous. The enlarged cross-section through the beam shows the applied loads.

Torsional effects may be treated for simplicity in terms of warping of the cross-section and ignoring the pure torsional resistance. This may be treated by considering equal and opposite transverse forces in the flanges in equilibrium with the applied torsion (see Figure 24).

An additional criterion is that the connections between the SLIMFLOR beam and the columns should be capable of resisting the total torsional moment transferred to the beam.

The simplest way to determine the dimension z is to assume a value and check it against the following 'unity factor' condition:

$$\frac{M_x}{M_b} + \frac{M_y}{M_{cy}} \leq 1.0$$

where:

M_x = applied moment about the x-x axis

M_b = buckling resistance moment capacity about the x-x axis (determined using BS 5950: Part 1 and the Section 4.1.2 of this publication)
Note, $M_b = M_c = S_x \cdot p_y$ when $I_y > I_x$

M_y = applied transverse moment to the top flange of the UC about the y-y axis. This is treated by considering the torsional moment as two opposing forces in the flanges.

M_{cy} = moment capacity of the top flange (UC) about the y-y axis (= $TB^2 p_y/4$).

The formulae required to evaluate the above expression are:

Applied moment, M_x

Factored loading

$$W_1 = \left[(w_{pc} \cdot \gamma_{fd}) + (w_c \cdot \gamma_{fc}) \right] \left[\frac{L_2}{2} \cdot z \right]$$

$$W_2 = w_s \, \gamma_{fd} \, L_1 . L_2$$

where:
w_{pc} = weight of pc units
w_c = construction load
w_s = self weight of steel
γ_{fd} = load factors for dead loads (1.4)
γ_{fc} = load factors for imposed loads (1.6)

Applied moment about the x-x axis $M_x = M_1 + M_2$

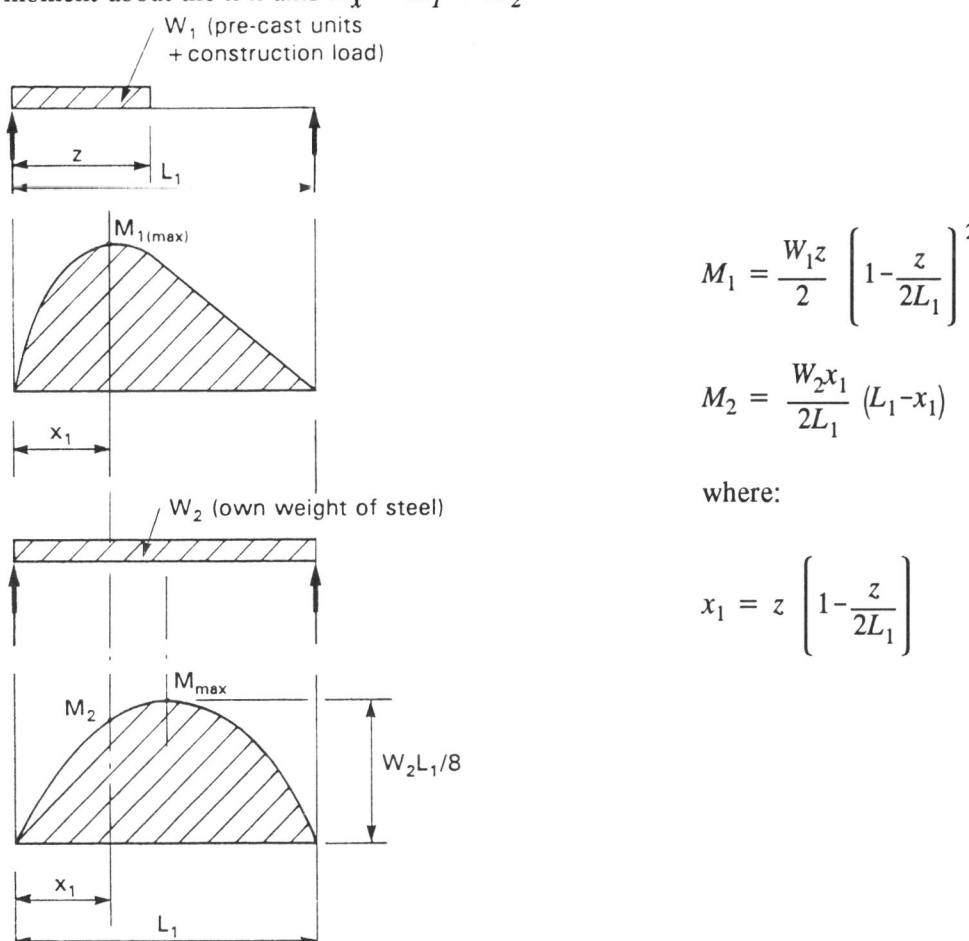

$$M_1 = \frac{W_1 z}{2} \left(1 - \frac{z}{2L_1} \right)^2$$

$$M_2 = \frac{W_2 x_1}{2L_1} (L_1 - x_1)$$

where:

$$x_1 = z \left(1 - \frac{z}{2L_1} \right)$$

Figure 25 *Bending moment diagrams for a partially loaded bay and self weight of steel members*

Transverse moment, M_y

Equivalent horizontal force, F, due to torsional effects on section.

$$F = \frac{W_1 e}{h}$$

where: $h \approx D - T/2$
e = eccentricity of load W, measured from the centre line of the UC section.

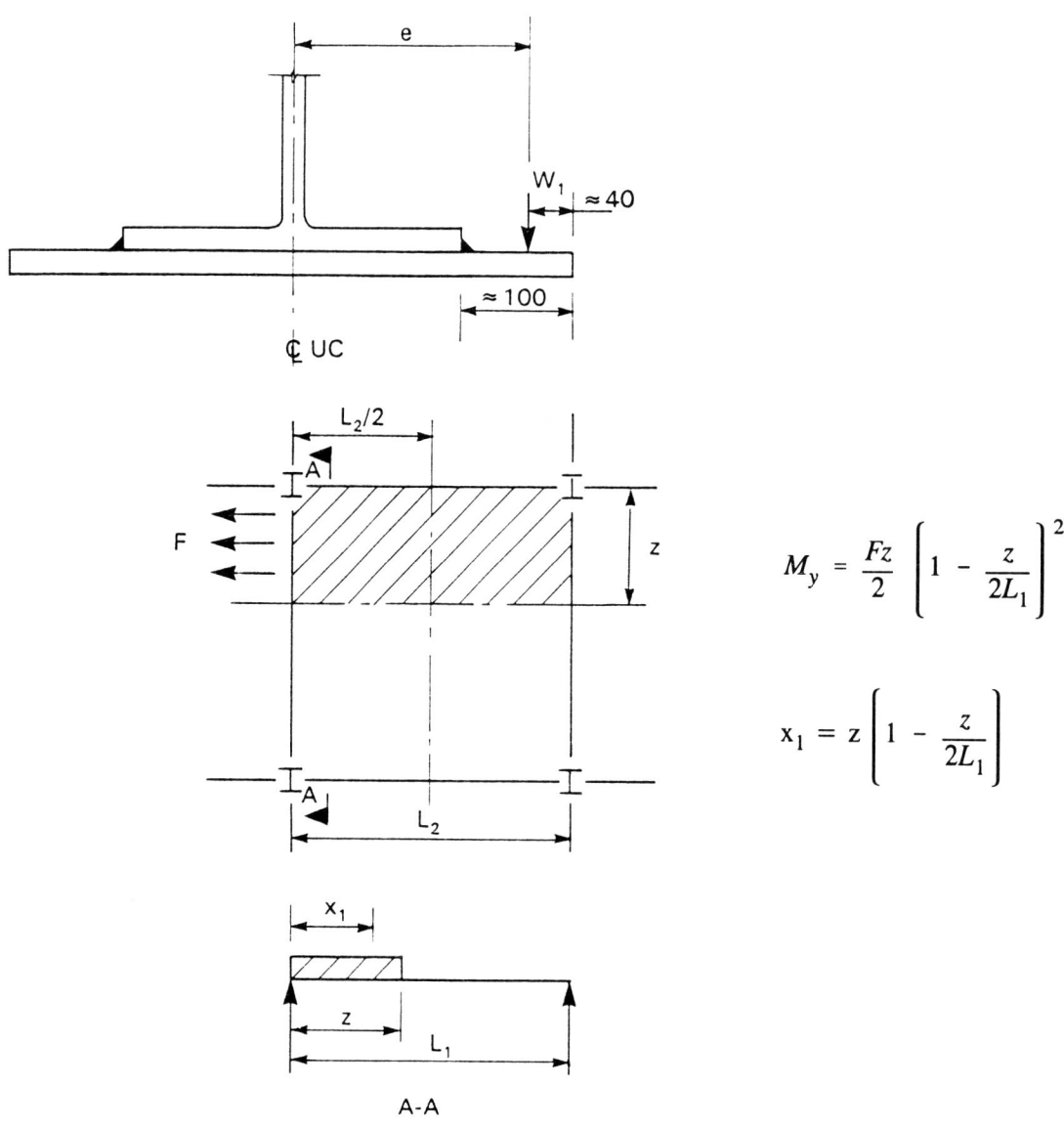

$$M_y = \frac{Fz}{2}\left(1 - \frac{z}{2L_1}\right)^2$$

$$x_1 = z\left(1 - \frac{z}{2L_1}\right)$$

Figure 26 *Diagrams showing the horizontal force, F generated from the out of balance load, W_1*

4.1.6 Summary of design procedures for the construction stage

1. Check the design of the SLIMFLOR beam where the beam is subject to the worst combination of bending and torsional effects ($z = L_1$ in the above equation). The lateral torsional buckling and torsion check is determined by satisfying the 'unity factor'.

2. If the 'unity factor' is exceeded, determine the value of z for acceptable design. This is the extent along the beam that the pc units can be placed before adjacent units in other bays have to be erected so that balanced loading can be restored.

   Clearly, if the units are placed symmetrically in a sequential pattern the torsional check need not be made.

A fully worked example illustrating the above procedures and the rigorous analysis using the SCI publication[6] are shown in Appendix A. These calculations for the combination of torsion and lateral torsional buckling design during construction gave the following results:

(i) Internal bay fully loaded:
   Rigorous analysis: unity factor = 1.00 (Worked example No. 3)
   Simplified analysis: unity factor = 1.31 (Worked example No. 3)

(ii) Internal bay partially loaded:
   In order to achieve a unity factor of 1.0 using the simplified approach, limit Z to 5 m.

The nature of construction stage loading is unpredictable especially when dealing with LTB combined with torsion. If the rigorous analysis is compared with the simplified approach it is apparent that a healthy degree of safety is incorporated into the simplified approach.

4.1.7 Vertical shear capacity

The vertical shear capacity, P_v, of the steel member is determined using BS 5950 : Part 1[1].

$$P_v = 0.6\, p_y . A_v$$

where A_v is the shear area taken as $t_w . D$.

Vertical shear can influence the moment capacity of the beam. This occurs where high shear and moment co-exist at the same position within the span. Simply supported beams with one or two point loads are good examples of where this happens. This case is not covered by this publication, as the scope of this design guide is restricted to simply supported beams with uniform loading only. For this case any influence the shear might have on moment capacity is considered as minimal.

4.1.8 Serviceability stresses

See Section 4.3.4.

4.2 Semi-Composite beam

Design assumptions

(a) The SLIMFLOR beam (see Figure 27) is considered as laterally unrestrained for the construction stage.

(b) The SLIMFLOR beam is considered as laterally restrained for the imposed loading condition (See 4.1.1(b)).

(c) Out-of-balance loads do not need to be considered for the imposed loading condition.

4.2.1 Section Classification

This is dealt with in a manner similar to the non-composite beam. However, the concrete that surrounds the section will provide additional support to the compression elements.

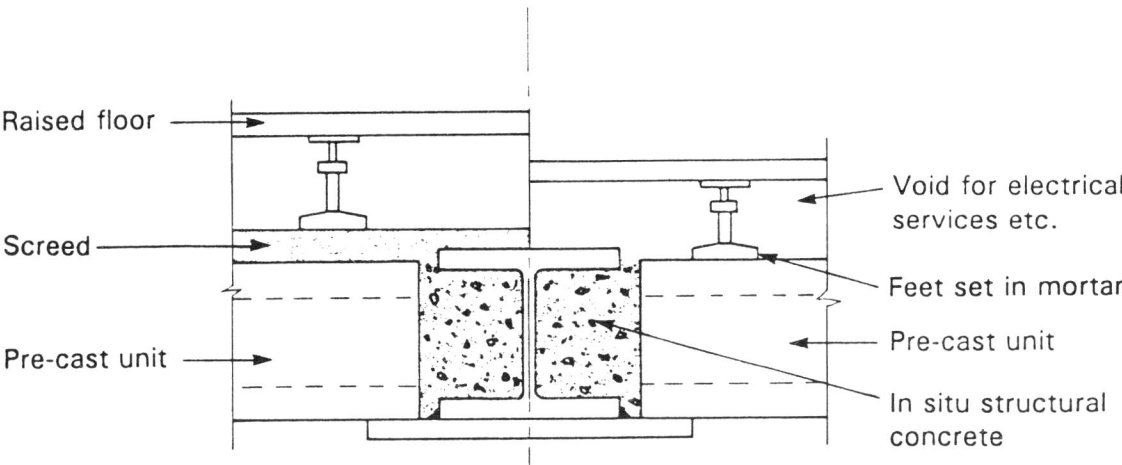

Figure 27 *Typical cross-section through the semi-composite beam*

4.2.2 Moment capacity

In the construction stage, LTB and pure torsion are treated in the same manner as the non-composite beam. The concrete that surrounds the beam is primarily used for stiffness purposes. However, it is assumed to provide adequate lateral restraint to the member for imposed loads at the ultimate limit state. Enhancement of the moment capacity of the beam by combining the surrounding concrete with the steel member is difficult to justify unless additional shear connection is provided. It is for these reasons that the composite action is neglected at the ultimate limit state thereby simplifying the design procedures. The following describes the method of deriving the equation for moment capacity (steel member only) using plastic analysis for the design of the cross-section.

Figure 28(a) shows the position of the plastic neutral axis (pna) in the web, at a distance y_p below the centre line of the Universal Column. To analyse the moment capacity from this diagram would involve some tedious calculations to obtain the moment capacity of the section. In order to simplify the calculation, Figure 28 (b) and (c) show a standard method of rearranging the rectangular stress blocks, so that the moment capacity may be easily obtained.

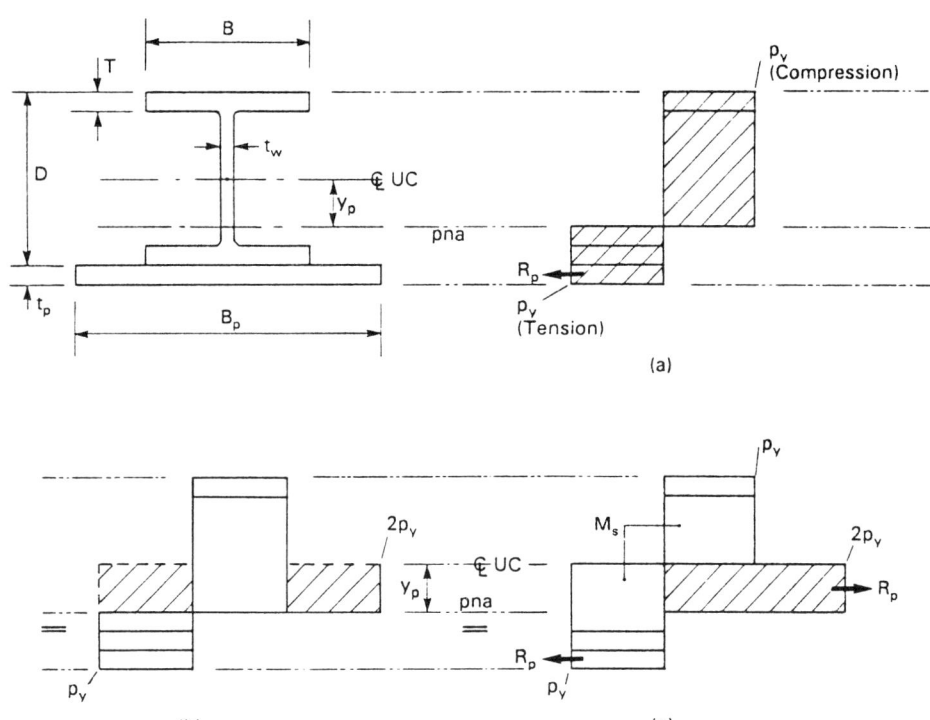

Figure 28 *Ultimate stress distribution through the "SLIMFLOR" beam*

From Figure 28 (c) y_p can be found from:

$$R_p = 2p_y t_w y_p$$
$$\therefore y_p = \frac{R_p}{2p_y t_w}$$

Moments are now taken about the centre line of the UC to find M_c.

$$M_c = M_s + R_p \left(\frac{D}{2} + \frac{t_p}{2}\right) - R_p \frac{y_p}{2}$$

but $y_p = \dfrac{R_p}{2p_y t_w}$

$$M_c = M_s + \frac{R_p}{2}(D + t_p) - \frac{R_p^2}{4p_y t_w}$$

where:

$M_s = S_x \cdot p_y$
$R_p = A_p \cdot p_y$
$A_p = $ Area of flange plate $(B_p \cdot t_p)$.

For other positions of pna, M_c can be calculated using the formulae given in Appendix B.

4.3 Composite construction

Design assumptions

Design assumptions as the semi-composite beam (see Section 4.2).

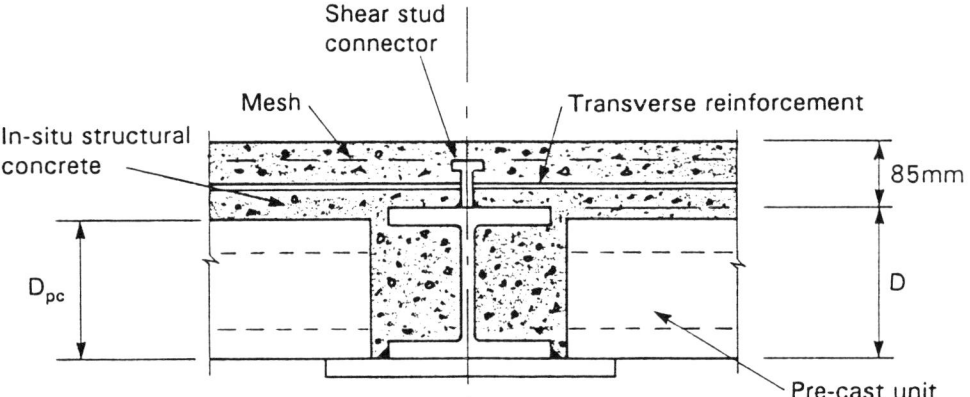

Figure 29 *Typical cross-section through the composite beam*

4.3.1 Section classification

See Section 4.1 for section classification.

Effective width of compression flange

The effective width of the compression flange in a "T beam" is limited by the influence of shear lag. For a wide concrete flange only a certain proportion of that flange can be mobilised to act with the beam. The concrete stress (maximum) starting from the centre line of the beam will gradually reduce at points remote from that centre line. The effective width, B_e, will vary for ultimate and serviceability limit states but a compromise value (Figure 30) of Span/4 for internal beams has been made in BS 5950: Part 3: Section 3.1[2]. The value of B_e as shown in Figure 30 should not exceed the distance between beam centres.

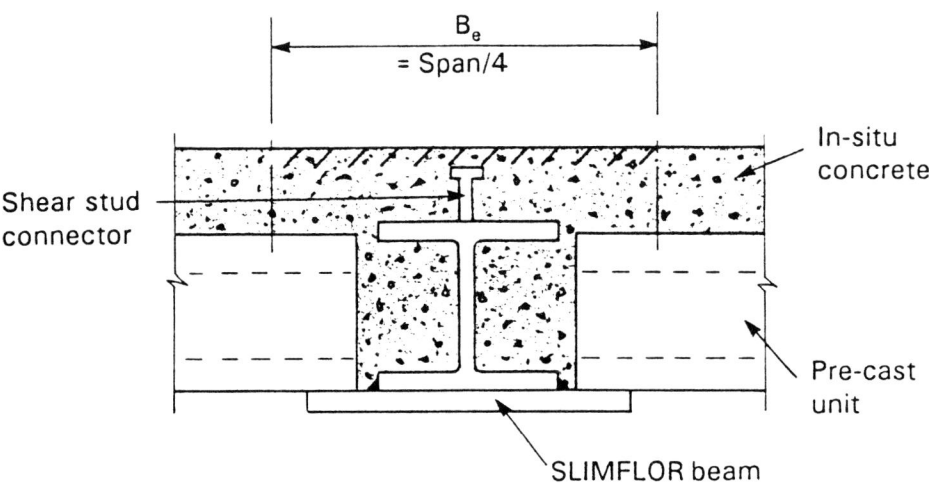

Figure 30 *Typical cross-section through the composite beam showing the effective width of the compression flange*

Modular ratio

The modular ratio, α_e, is defined as the ratio of the elastic moduli of steel to concrete which is generally expressed as:

$$\alpha_e = E_s/E_c$$

and is used for determining the elastic properties of a composite section. A value of 205 kN/mm^2 is used for the elastic section modulus of steel (see BS 5950: Part 1). The short term elastic modulus for normal weight concrete is in the range of 25 to 30 kN/mm^2 depending on concrete grade. As concrete strengths increase with age the elastic modulus may become 2 to 3 kN/mm^2 higher. This would give actual values for α_e (short term) in the range of 6 to 7.

The elastic modulus for lightweight concrete is lower than that of normal weight concrete. For the short term, values can be expected to be in the range of 10 kN/mm^2 to 11.5 kN/mm^2 for lightweight concrete. Modular ratios for long term loading are creep dependent which will have an effect on the elastic modulus of the concrete. This influence can lead to values of α_e for long term loading of 18 to 21 for normal weight concrete and 25 to 29 for lightweight concrete.

Table 3 is an extract from BS 5950: Part 3: Section 3.1 showing modular ratios for long and short term loading.

Table 3 *Modular Ratios*

Type of concrete	Modular ratio for short-term loading α_s	Modular ratio for long-term loading α_ℓ
Normal weight	6	18
Lightweight	10	25

For buildings of normal usage the modular ratio should be assumed to be ⅔ short term and ⅓ long term values. This gives values of 10 and 15 for normal and lightweight concrete respectively for use in imposed load calculations.

4.3.2 Moment capacity, M_c

The moment capacity for composite slim floors is dependent on the degree of shear connection between the concrete and steel beam. The simplest way to illustrate full and partial shear connection is to use an example. Figure 31 shows a typical cross-section through a composite "SLIMFLOR" beam with full shear connection. Also shown is the plastic stress distribution across the section. (See Appendix B - Case 2FS for full shear connection.)

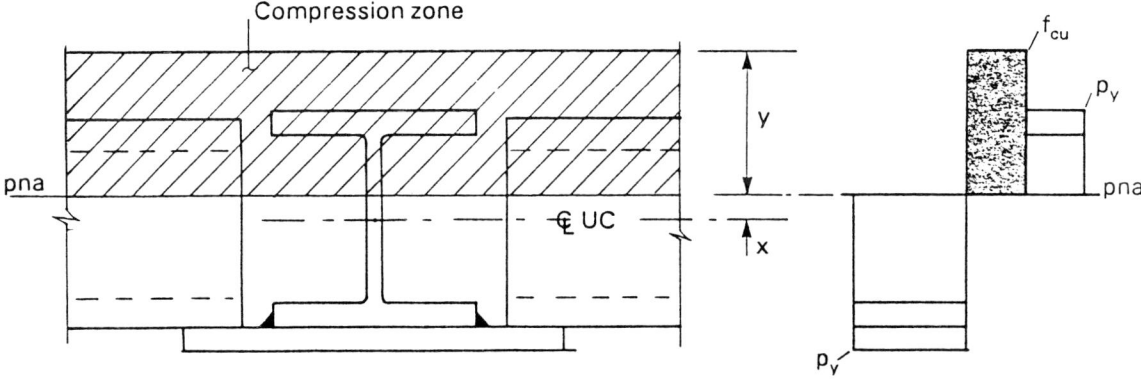

Figure 31 *Typical cross-section through composite slab with full shear connection*

Full shear connection

Full shear connection occurs where the number of shear connectors provided is greater than that required to develop full resistance of the concrete or steel member. This will generate the maximum moment capacity of the cross-section. In the majority of cases, the concrete resistance, R_c is less than the resistance of the steel member, $(R_s + R_p)$

i.e. $R_c < (R_s + R_p)$

where: $R_c = 0.45 f_{cu} B_e . y$
$R_s = A.p_y$
$R_p = A_p.p_y$

In the above equations the only unknown is y. This is the distance to the pna from the top of the slab. The appropriate formulae for x and y can be found in Appendix B. The procedure is to calculate x (pna to the centre line of the UC) by trial and error. Once x has been established then y can be evaluated.

For example when the pna is located in the position shown in Figure 31 the equations for x and y are:

$$x = \frac{0.45 f_{cu} B_e (D_s + D_{pc} - D/2) - R_p}{0.45 f_{cu} B_e + 2p_y t_w}$$

$$y = D_s + D_{pc} - D/2 - x$$

Once y has been determined then R_c can be evaluated.

In cases where the pna lies within the pc units, higher forces may be generated in the shear connectors leading potentially to an overload in longitudinal shear. Therefore, to account for this effect, the number of shear connectors is determined for the maximum possible force developed in the concrete (including the pc units), or in the steel section. Conversely, when considering the moment capacity it is prudent to assume that this bonding of the concrete no longer exists. In these circumstances a conservative approach has to be adopted. Otherwise, the degree of shear connection could be lower than the minimum required value.

For convenience the characteristic strength of the pc units has been assumed to be the same as the in situ concrete. The characteristic strength of the pc units is in the region of 60 N/mm^2 which is much higher than the in situ concrete say, 30 N/mm^2. As the full cross section of the units has been assumed to act with the in situ concrete this difference in strength would appear to be reasonable due to the voids in the pc unit cross-section, i.e. the voids compensate for the higher characteristic strength of the pc unit. The procedure for full shear connection is:

(a) assume position of pna

(b) calculate distance to pna, x, using appropriate case formula showing in Appendix B

(c) check if pna is where assumed. If not, assume another position of pna and repeat until assumed position of x corresponds to the calculation

(d) calculate y using appropriate case formulae shown in Appendix B

(e) calculate force in concrete, R_c

(f) calculate the force $(R_s + R_p)$

(g) the shear connection force equals the lesser of R_c and $(R_s + R_p)$

(h) the calculation of the moment capacity for full shear connection is not required for the reasons given above.

Partial Shear Connection

Partial shear connection is used when there is an excess of moment capacity compared to the applied factored moment. When this situation occurs BS 5950 : Part 3 : Section 3.1[(2)] allows a reduction (to a maximum of 40%) of the number of shear connectors needed for full shear connection.

The principle is that the number of studs are reduced so that the force R_q (Figure 32) is sufficient to provide a similar moment capacity when compared to the applied factored moment.

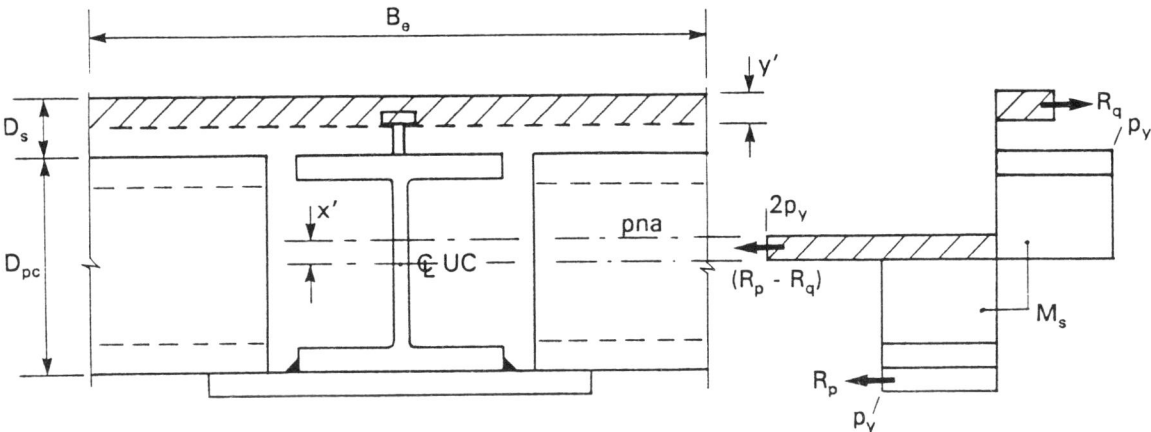

Figure 32 *Typical cross-section through composite slab for partial shear connection*

The formulae for moment capacity can be found in Appendix B. For this example the moment capacity, M_c, is given by the following expression.

$$M_c = M_s + R_q \left[D_s - \frac{D}{2} - \frac{R_q}{0.9 f_{cu} B_c} \right] + \frac{R_p}{2} (D + t_p) - \frac{(R_q - R_p)^2}{4 p_y t_w}$$

where: $M_s = S_x \cdot p_y$

In the equation for M_c the only unknown is R_q.

The procedure for partial shear connection is:

(a) assume a degree of shear connection, e.g. 40% then R_q = 40% of the longitudinal shear transfer (from g above)

(b) assume position of pna, e.g. in bottom of the flange

(c) calculate x using the appropriate formula in Appendix B

(d) check whether pna is where assumed, if not, assume another position and repeat

(e) calculate y using appropriate formula in Appendix B

(f) find maximum limit to y' (y'_{crit}) using formula. Figure 33 limits the depth of in situ concrete in compression below the top flange. This practical limit of 50 mm is to discourage the use of shallow depth planks combined with in situ concrete. If there was no limit on the depth $D^{(1)}$, shown in Figure 33, the system could be over reliant on the in situ concrete. This would probably require additional shear connectors and transverse reinforcement to mobilise the compressive forces around the top flange (UC). As previously mentioned it is preferred where possible to keep the depths D and D_{pc} similar, thereby overcoming this concern.

(g) check that $y' \leq y'_{crit}$. If not so, redesign structural section

(h) calculate moment capacity, M_c from appropriate formula shown in Appendix B

(i) check that applied factored bending moment $\leq M_c$. If not so, repeat procedure with a higher degree of shear connection until condition satisfied or redesign structural section and repeat procedures.

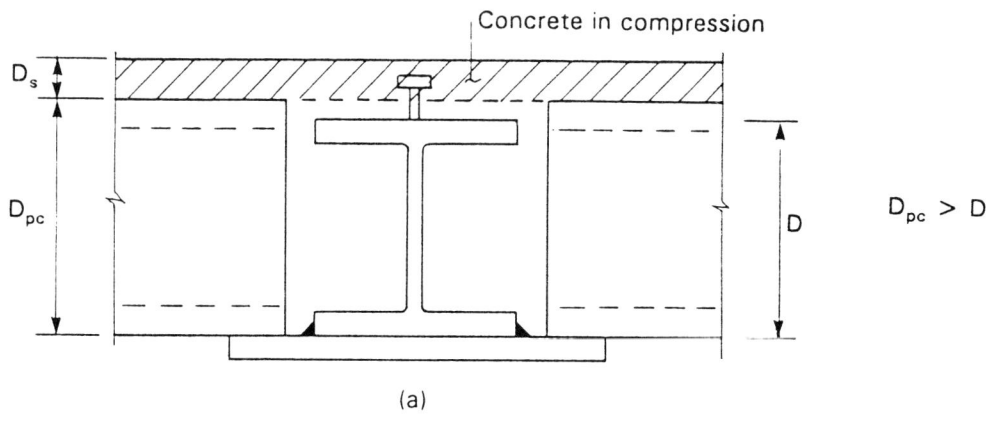

(a)

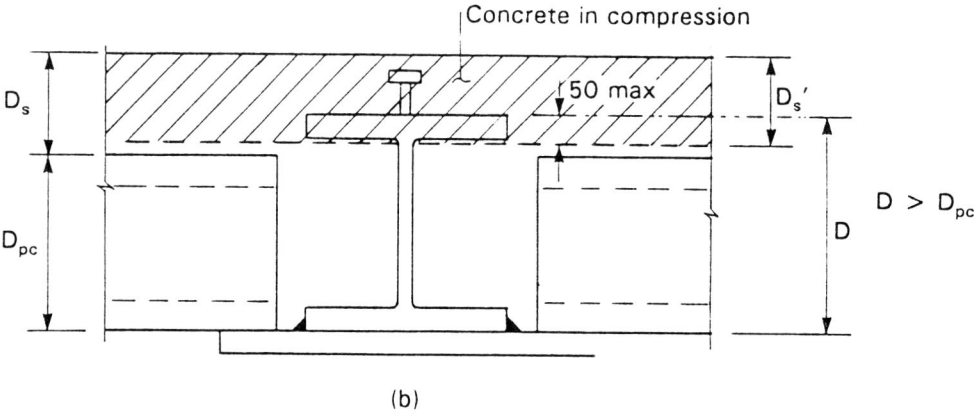

(b)

Figure 33 *Limitations on in situ concrete depth in compression*

4.3.3 Deflection

The reason for slim floor construction is, as the name implies, to provide the minimum in floor depth construction. Shallow beam depths with relatively long spans are sensitive to deflection. This is one reason to span the shorter distance with the supporting beam and allowing the pc units to provide the greater span.

The deflection limits for beams are given in BS 5950: Part 1. The imposed load limit is span/360 based on unfactored loads.

Pre-Cambering

The implication of the final deflection due to dead loads may be considered as important. This deflection is the accumulation of deflections due to its self weight, pc units, in situ concrete and reinforcement. Where this deflection is considered excessive, the beam may be pre-cambered to nullify these dead load deflections. In addition, the welding of the flange plate to the UC section will cause some degree of pre-camber. An alternative to pre-cambering may be to increase the beam size but this option would depend on costs. To define or quantify excessive deflection is difficult because the requirements will vary from one building to another. However, as an approximation, a deflection limit of span/200 for total loading (dead and imposed loads) will provide an indication as when to consider pre-cambering.

Partial shear connection

As a consequence of the methods used for determining the moment capacity an increase in vertical deflection will have to be allowed for under serviceability loads.

Partial shear connection design results in a greater degree of slip occurring in the shear connectors. The result of this action is an increase in vertical deflection. This is given by the following expression for unpropped construction:

$$\delta = \delta_c + 0.3 \left[1 - \frac{N_a}{N_p}\right] (\delta_s - \delta_c)$$

where: δ_c = the deflection of the composite beam under imposed load

δ_s = the deflection for the steel beam acting alone under imposed load

N_a/N_p = the degree of shear connection.

4.3.4 Serviceability stresses

Stresses at the serviceability limit state are calculated to ensure that under working loads no permanent deformations can be formed in the steel member.

The stress in the extreme fibre of the steel beam should not exceed the design strength, p_y and the stress in the concrete flange should not exceed $0.5 f_{cu}$.

The transverse moments in the bottom flange plate have to be combined with the longitudinal stresses and compared to the design strength of the steel member, p_y.

No account is taken of the effect of slip on these stresses and the associated forces on the shear connectors at the serviceability limit state. Serviceability stresses are included for the three worked examples shown in Appendix A.

4.3.5 Transverse reinforcement

Transverse reinforcement is used to ensure a smooth transfer of the longitudinal force at the ultimate limit state (via the shear connectors) into the slab without splitting the concrete. Potential shear failure planes through the slab lie on either side of the shear connectors as shown in Figure 34. For further information see References (7) and (8).

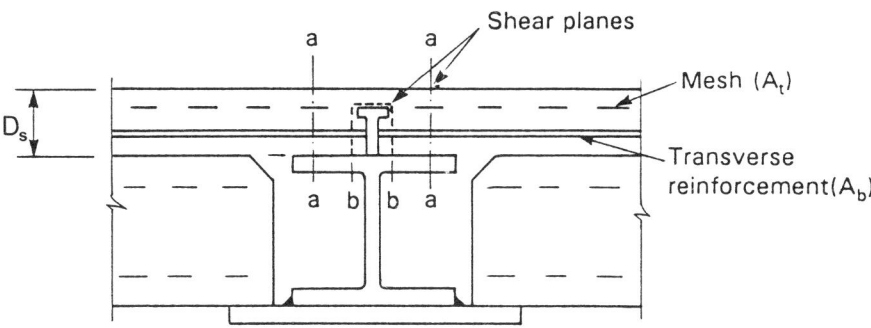

Figure 34 *Assumed shear failure planes at the ultimate limit state*

The shear resistance per unit length of each plane along the beam is given by:

$$v_r = 0.8 A_{sv} f_y + 0.03 \eta A_{cv} f_{cu}$$

but $v_r \leq 0.8 \eta A_{cv} \sqrt{f_{cu}}$

where:

f_{cu} = characteristic cube strength of the concrete in N/mm², but not greater than 40 N/mm²

η = 1.0 for normal weight concrete

η = 0.8 for lightweight concrete

A_{cv} = mean cross-sectional area, per unit length of the beam, of the concrete shear surface under consideration

A_{sv} = cross-sectional area per unit length of the beam, of the combined top and bottom reinforcement crossing the shear surface, see Figure 34 and Table 4.

Table 4 *Reinforcement area and lengths of shear plane for Figure 34*

Shear plane	A_{sv}	Length of shear plane	
		one plane	two planes
a-a	$A_b + A_t$	D_s (i)	$2D_s$
b-b	$2A_b$	-	see below (ii) & (iii)

(i) D_s is the depth of concrete above the top flange of the UC

(ii) For one row of shear connectors the length of shear plane = $2h$ + head diameter of stud

(iii) For two rows of shear connectors the length of shear plane = $2h + S_t$ + head diameter of stud

where: h = height of stud
S_t = transverse spacing centre-to-centre of the studs

The longitudinal shear force, per unit length, v to be resisted can be obtained from the spacing of the shear connectors.

$$v = \frac{NQ}{s}$$

where: N = number of shear connectors in a group
Q = shear connector value for positive moments
s = longitudinal spacing of the shear connectors

4.3.6 Shear connector design

The most popular size of shear stud connector available is the 19 mm diameter × 100 mm high (95 mm after welding). However, to keep the in situ concrete to a minimum thickness over the top flange (UC) it is proposed to use the 19 diameter × 75 high (70 mm after welding) stud. Allowing for 15 mm cover the slab depth over the flange becomes 85 mm (minimum depth). Any unnecessary increase in the concrete depth will obviously add further to the dead loads and produce greater construction depths. The studs are welded to the top flange of the beam in the works prior to sending it out to site. This procedure has the advantage over site welding in avoiding the weather conditions which can interfere with the welding procedures. Table 5 gives the characteristic resistance values for headed studs in normal weight concrete. For the use of lightweight concrete the characteristic resistance should be taken as 90% of the value given in Table 5. Where the dead loads are considered high and deflections are to be kept to a minimum, the use of lightweight concrete may be a suitable alternative to normal weight concrete.

Table 5 *Characteristic resistances (kN) of headed stud shear connectors from BS 5950: Part 3*

Dimensions of Stud Shear Connectors (mm)			Characteristic strength of concrete (N/mm^2)			
Diameter	Nominal Height	As-welded Height	25	30	35	≥ 40
25	100	95	146	154	161	168
22	100	95	119	236	132	139
19	100	95	95	100	104	109
19	75	70	82	87	91	96
16	75	70	70	74	78	82
13	65	60	44	47	49	52

Notes:
1. For connectors of height greater than tabulated, use the values for the greatest height tabulated.
2. Data for normal weight concrete.
3. Design strength = 0.8 × characteristic strength in positive moment regions.

Shear connector spacing

The maximum longitudinal spacing between the shear connectors should not exceed 600 mm or 4 times the slab depth. In this instance the slab depth is taken as the lesser of depth of in situ concrete above the top flange (UC) or the pc unit. Figure 35 shows a plan of the beam top flange with the limiting dimensions for stud spacing.

The minimum longitudinal spacing is 5 times the stud diameter, d_s, and the minimum transverse spacing is 4 times the stud diameter (for studs in pairs).

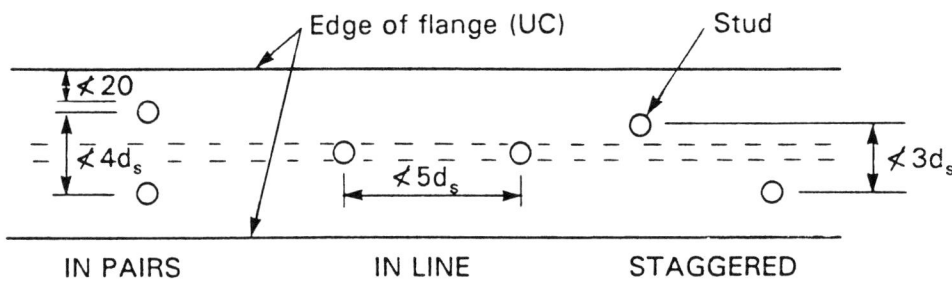

Figure 35 *Requirements for dimensional spacing of shear stud connectors*

To prevent "burn through", the diameter of the stud should not exceed 2.5 times the thickness of the beam flange (not normally a problem with UC sections).

4.3.7 Natural frequency of the beam

The "SLIMFLOR" system provides for shallow construction depths with relatively high dead loads. The combination of these two factors can lead to the floor structure being susceptible to vibration. The "natural frequency" of the floor or beams is the parameter commonly associated with this effect. This design check is normally considered where the spans are long but in view of these higher dead loads it would be prudent to check all beams, irrespective of the span. For the purposes of design, a lower limit of 4 Hz (cycles per second) has been found to be adequate for the natural frequency of the beam. The loading used for the calculation comprises dead loads and 10% of the imposed load. In this instance partitions are not included because they produce a dampening effect on the floor structure.

To determine the natural frequency of the floor beam the following approximate expression may be used:

$$f = \frac{18}{\sqrt{\delta_{sw}}}$$

δ_{sw} is the deflection of the beam (non composite or composite, as appropriate) subject to the self weight of the floor including 10% of the imposed loads (but not including the partition loads, see above). The effects of partial shear connection need not be made. A 10% reduction in deflection may be made for composite sections to account for the increased dynamic stiffness of the beam.

Where a more exacting design is required reference should be made to the SCI publication *Design guide on the vibration of floors*[9]. Using this design guide it may be possible to justify the use of a lower value for the natural frequency.

4.4 Edge beams

This section should be read in conjunction with Sections 3.2 and 4.1.

The first consideration for the edge beam design is to determine what form of construction would be permitted. The selected form of construction for the edge beam will dictate the design concept. For example, if a downstand beam (see Figure 13) will incorporate the cladding details a more traditional design approach can be adopted. The pc units can rest on the top flange of the beam and the vertical loads may be assumed to act through the shear centre of the beam. This has the advantage of eliminating the torsional effects on the beam.

Edge beams which have eccentric loads will have to cater for the torsional effects in a similar manner to the procedures given for an internal beam. Alternatively, the loads can be shown to act through the shear centre of the section or the external cladding loads can balance these eccentric forces. Where a net torsional moment exists this will have to be combined with the lateral torsional buckling of the section unless it can be shown that the compression flange of the section is otherwise restrained.

4.5 Connections

The beams are assumed to be simply supported which means that the end plates should be of a flexible nature. The use of a partial depth end plate (end plate welded to the beam web only) would provide the flexibility for the connection to be considered "simple". However, the stages of construction have to allow for the torsion of the beams (due to out of balance loads) which makes the partial depth end plate unsuitable in this instance. For these reasons, it is recommended to use a full depth welded end plate. This type of end plate detail should provide a connection suitable of withstanding beam rotations. In addition, the full depth end plate will give extra stability to the frame in the erection stage.

An end plate which is welded to the flanges and web of the beam can be considerably stiffer than a partial depth end plate. To ensure that the connection remains "flexible"; Reference 10 recommends that a limit is placed on the plate thickness and bolt centres. Typical values are:

End plate thickness mm	10 (max)	8
Bolt cross centres mm	140	90 (min)

Based on an extensive series of fire tests on SLIMFLOR construction, it is recommended that the minimum size of fillet welds is 8 mm. The weld design procedures are given in Appendix A, worked example No.3.

Various arrangements exist for the tie to column connection. In the majority of cases the tie member will have small amounts of applied load. In those circumstances the design criterion is likely to be the requirements for robustness (see Section 4.6).

4.6 Robustness

This section is a brief explanation concerning the requirements for robustness of steel structures. Essentially, these requirements can be found in two documents:

(1) Building Regulations[4]
(2) BS 5950: Part 1: 1990[1]

The difference between these two documents is that Building Regulations only apply to structures which are more than four storeys high (including basements) or the span exceeds 9 m in a public building. Conversely, BS 5950: Part 1 considers all of the frame irrespective of the number of storeys and span. Compliance with these two documents will ensure that the structure has reached a prescribed minimum acceptable level of robustness, even in the most modest of structures. Basic requirements for robustness are given in Clause 1.01 of BS 5950: Part 1. The minimum tie forces to be resisted are stated in Clause 2.4.5 as 75 kN and 40 kN at floor and roof levels respectively. Structures greater than four storeys in height or spans which exceed 9 m (public buildings) must be appraised for their behaviour under collapse conditions. For structures over four storeys, three alternative design procedures are given in BS 5950: Part 1:

(1) Tying forces
(2) Localisation of damage
(3) Key elements

No guidance is given on measures for public buildings over 9 m span.

(1) Tying forces

This approach has a number of conditions which have to be satisfied for a robust design to be achieved. These conditions are:

(a) Sway resistance
(b) Tying: (i) General (ii) At the periphery
(c) Columns
(d) Integrity
(e) Floor units

(a) Sway resistance
Sway resistance may be achieved by the use of steel bracing, rigid joints, shear walls, stairwells and lift cores etc. These structural components should be sufficiently distributed through out the building, so that in each orthogonal direction, no substantial portion of the structural frame has to rely on a single plane of bracing.

(b) Tying
(i) General
The tying force requirements for internal and edge ties given in BS 5950: Part 1 are given by:

$0.5 w_f s_t L_a$ for internal ties

$0.25 w_f s_t L_a$ for edge ties

but not less than 75 kN for floors or 40 kN at roof level.

where: w_f = the total factored dead and imposed load

s_t = the mean transverse spacing of the ties

L_a = the greatest distance, in the direction of the tie, between adjacent lines of columns or other vertical supports.

Note, provided all the spans are equal and all the beams are used as ties, the above tying forces are equal to the ultimate vertical reaction at a beam support.

(ii) At the periphery
An additional requirement for structures greater than four storeys high is that ties which anchor columns at the periphery of a floor or roof should provide a minimum tying force equal to 1% of the factored axial load in the column. This effectively ensures positional restraint at the node points of the columns. The forces generated from the 1% of the factored axial load (as above) should not be less than the requirements given in (b)(i) above.

Note, see also Clause 4.7.1.2.

(c) Columns
Column splices should be capable of resisting a tensile force which is dictated by the loading on the floor below. This tensile force equals two-thirds of the factored vertical load applied to the column from the next floor level below the splice. The exception is where the steel framework is of continuous construction in at least one direction. In this case the columns should be carried through at each beam-to-column connection.

(d) Integrity
Where a beam carries a column it should be checked, together with the member which supports it for localisation of damage (see (2) below).

(e) Floor units
Pre-cast concrete floor units should be adequately tied either to each other over the support, or directly to the supporting member. Pre-cast units tied together can act in two ways. Firstly, the ties help to maintain the integrity of the floor following the removal of the primary supporting element and thereby limit the damaging effect of the units falling as debris onto the floors below. Secondly, where the pc units are adequately tied the diaphragm action of the floor is developed. This tying action will provide an improved distribution of resistance to lateral loads under normal design conditions.

Figure 36 shows a typical detail on tying pc units used in slim floor construction.

(2) Localisation of damage

When the requirement for tying forces are not met or as an alternative to "tying forces" the building should be appraised in accordance with the requirements for the localisation of damage.

The provisions of BS 5950: Part 1 establishes whether at each storey in turn any single column or beam (including a beam carrying a column), could be removed without causing collapse of more than a limited portion of the building local to the member concerned.

This is based on the principle that structures should be inherently capable of limiting the spread of local failure regardless of the cause.

BS 5950: Part 1 does not give the value for the permitted extent of the local failure. The intention is to refer to the Building Regulations[8] for the permitted extent of the damage which is currently quantified as not exceeding:

The immediately adjacent storeys or beyond an area within those storeys of:
(a) 70 m² or
(b) 15% of the area of the storey whichever is less.

If this cannot be satisfied then the member must be designed to a 'key element' (see (3)).

These requirements would appear to be restrictive for longer span steel framed buildings. As a result, it is understandable why beams which support areas of floor greater than 70 m² are often designed as a key element.

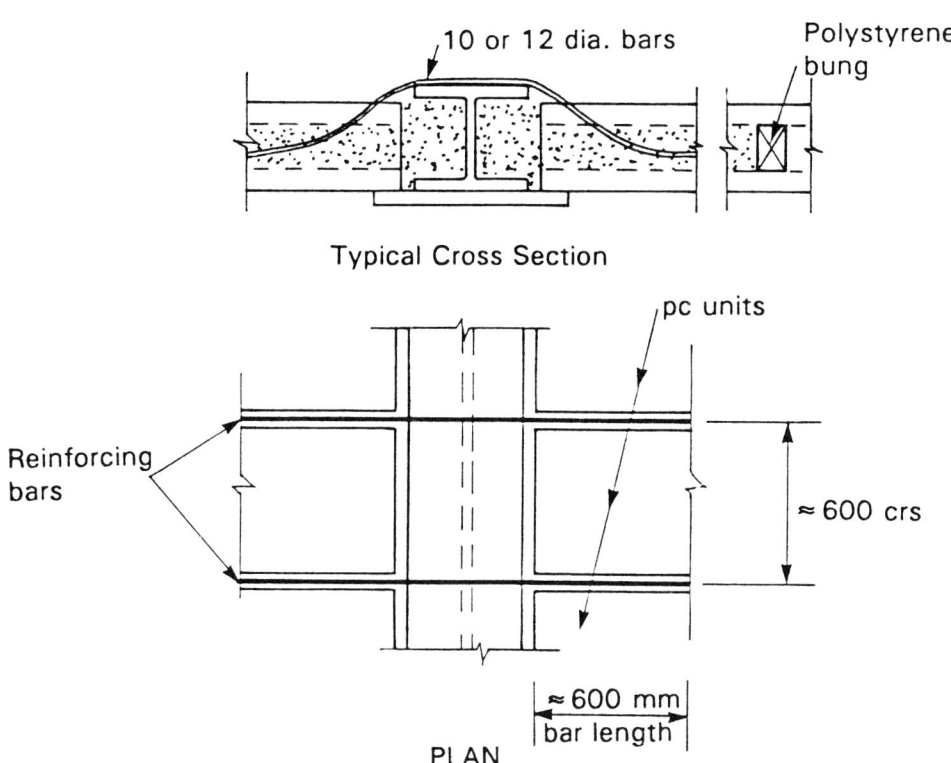

Figure 36 *Typical tying detail*

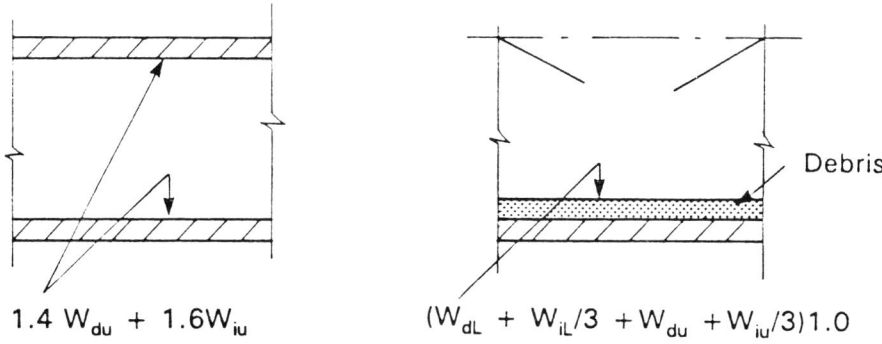

Figure 37 *Cross-section through multi-storey building showing the collapse of the upper storey*

Figure 37 shows a section through a multi-storey frame where the dead and imposed loading on the upper floor are denoted as W_{du} and W_{iu} whilst those on the lower floor as W_{dL} and W_{iL} respectively. The assumption is that in the event of an abnormal sequence of loading, the upper storey collapses and the resulting debris impacts on the floor below.

Progressive collapse of the floors down through the structure is prevented by designing the lower floor to resist an ultimate load given by:

$$\left[W_{dL} + \frac{W_{iL}}{3} + W_{du} + \frac{W_{iu}}{3} \right] \times 1.05$$

When the imposed load is of a permanent nature or the building is used predominantly for storage, the full imposed load should be used.

Clearly, the above equation does not include for the dynamic impact effects of the debris. The assumption is that the units are adequately tied therefore, at the onset of collapse the floor units behave with catenary action which has the effect of reducing the impact velocities. Where the pc units are not adequately tied over the supports (say, type 1 "SLIMFLOR" beam) then dynamic effects should be considered. As a guide, the debris loading could be increased to twice that of the static load on the upper floor. This type of appraisal can only be deemed satisfactory if failure of a beam results in the collapse of less than 70 m^2 of the floor area.

(3) Key elements

A member is essentially categorised as a "key element" when the tying force and localisation of damage routes do not comply with the relevant clauses of BS 5950: Part 1[1]. Elements in these circumstances for which significant proportions of a structure rely on for stability and support can be designed as a "key element". For this condition the member is designed to resist abnormal loadings which may otherwise render it ineffective.

In BS 5950: Part 1[1] no specific guidance is given on the magnitude of abnormal loads which should be applied. However, the recommendations for the reduced imposed loading and load factors given for localisation of damage (see 2 above) may be considered for an abnormal occurrence. Generally, the accepted value for such loading (blast pressure) is 34 kN/m^2 as stated in the Building Regulations[4].

A practical situation where these above recommendations could be used are for columns in multi-storey buildings. In this instance columns are sometimes categorised as a "key element" due to the large areas of floor which they support. The column is subjected to a lateral blast pressure of 34 kN/m^2 which is considered acting over the projected area of the column in conjunction with the dead and imposed loads. The design loading can therefore be expressed using BS 5950: Part 1[1] and the Building Regulations[4] as:

$$(W_d + \frac{W_i}{3} + 34) \times 1.05 \quad (kN/m^2)$$

In addition to the above BS 5950: Part 1[1] establishes that the accidental loads should be applied to members from appropriate directions together with the reactions from other building components attached to the members which are subjected to the same loading but limited to the ultimate strength of these components or their connections.

5. FIRE RESISTANCE

The fire resistance of SLIMFLOR beams has been investigated both experimentally, by a number of fire resistance tests, and analytically using computer modelling techniques. As a result of these studies recommendations have been produced for up to 60 minutes fire resistance without the use of any applied fire protection. For more than 60 minutes fire resistance additional fire protection is generally recommended.

The fire resistance of a slim floor beam is inherently good because most of the section is shielded from the fire by the floor units. All the slim floor systems described in this guide achieve 30 minutes fire resistance without any applied fire protection or any loading restrictions. An exposed universal beam would have to be at least a 610 × 305 × 179 to have similar performance. For 60 minutes fire resistance, with minor exceptions, Type 1 and Type 2 systems can be used unprotected with little restriction on loading. However, the composite system, Type 3, may require additional protection or may have loading restrictions. For 90 minutes fire resistance applied fire protection is currently recommended.

5.1 Fire tests

The recommendations are based on the analysis of 8 fire resistance tests on loaded beams. These are summarised in Table 6. In tests 1 and 2 the floor units were built directly into the beam web and no supporting plate was used. In tests 3 to 8, 15 mm bottom plates were used together with 8 mm fillet welds.

Table 6 *Summary of fire resistance tests*

Test	Section UC size x kg/m	Type	Fire Resistance (mins)	Load Ratio	Plate (†) Temperature (°C)	Flange (†) Temperature (°C)
1	254 x 254 x 73	-	44	0.56	-	746
2	254 x 254 x 89	-	93	0.42	-	783
3	254 x 254 x 107	1	60	0.55	799	661
4	203 x 203 x 86	3	68	0.44	727	558
5	203 x 203 x 60	1	82*	0.51	812	691
6	254 x 254 x 73	2	78	0.47	778	578
7	152 x 152 x 30	1	75*	0.48	788	731
8	305 x 305 x 283	1	115*	0.17	728	411

Note: * Test discontinued before failure.
 † at 60 minutes.

For tests 7 and 8 it was decided to test a Type 1 design with the section only half filled with concrete as temperatures recorded would then be conservative when applied to a Type 2 or 3 design. In tests 3 and 5 sand infill was used which resulted in a lower fire resistance compared to concrete infill.

It can be seen from Table 6 that there is always an appreciable temperature difference between the plate and the bottom flange. This difference is very important and is one of the reasons why unprotected slim floor beams have such good fire resistance. It occurs due to an interface resistance between the two surfaces. Mathematical modelling has shown that the resistance is fairly constant and the temperatures can be predicted fairly accurately, with the exception of test 7. In test 7 the interface resistance between the 152 × 152 × 30 UC and the plate was appreciably lower than in the other tests. This has led to the adoption of a more cautious approach for all 152 × 152 sections.

The design procedure for slim floor beams described in this guide assumes that the precast units rest on the plate. All load transfer is therefore via the plate and the weld. The performance of the plate and weld was observed in the tests and was found to be satisfactory in that no appreciable transverse bending displacement occurred and no weld failures occurred during the normal test period. In two of the tests after the beam began to fail in overall bending the applied loads were modified. The loads causing overall bending were reduced and the loads causing the transverse bending on the plate were increased in an attempt to induce either a plate or weld failure. A weld failure was induced in one case but this occurred at a load far greater than could be expected in practice and at a temperature approaching 900°C. It was concluded that for up to 60 minutes fire resistance neither the unprotected plates nor the welds are critical to the design. A minimum size of 8 mm fillet welds is recommended. For 90 minutes fire resistance the performance of the plate and weld could be critical to the design of unprotected sections.

5.1.1 Analysis of fire tests

The fire resistance tests were analysed both structurally and thermally using programs developed by the SCI. The analyses were confirmed by British Steel using similar programs. The purpose of the analyses was to understand the observed performance in the tests and to develop a design procedure for any section in fire. In Figure 38, the calculated and observed deflections in one of the tests are compared. In Table 7 the measured and calculated temperatures for all the tests incorporating a bottom plate are compared and in Table 8 the moment capacities are compared. The moment capacities were calculated using the method described in BS 5950: Part 8, Appendix E [11], modified for slim floors. The plastic moment capacity of the section at elevated temperature is calculated knowing the temperature distribution and reduced strength of all the elements of the cross-section. The moment capacities have been calculated for the specific conditions and may differ from the recommended design values given in Table 11.

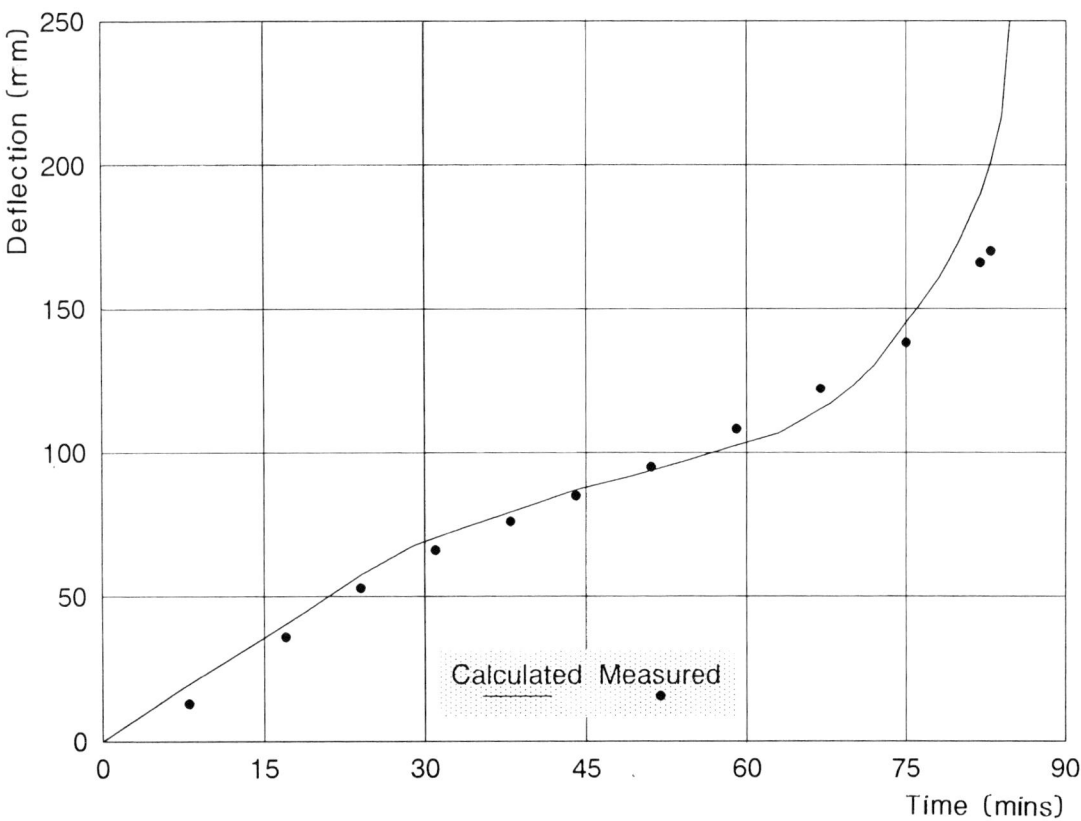

Figure 38 *Comparison of measured and calculated performance in Test 5*

Table 7 *Comparison of measured and calculated temperatures (°C)*

Test	Section	30 Minutes				60 Minutes			
		Test		Calculation		Test		Calculation	
		Plate	Flange	Plate	Flange	Plate	Flange	Plate	Flange
3	254 x 254 x 107	622	317	587	387	799	661	806	619
4	203 x 203 x 86	515	191	569	293	727	558	784	536
5	203 x 203 x 60	548	379	613	405	812	691	829	656
6	254 x 254 x 73	577	227	571	334	778	578	786	581
7	152 x 152 x 30	548	484	594	366	788	731	803	602
8	305 x 305 x 283	518	177	531	209	728	411	752	456

Table 8 *Comparison of moment capacities calculated using both measured and calculated temperatures*

Test	Section	M_c	30 Minutes		60 Minutes	
			Test	Calc.	Test	Calc.
3	254 x 254 x 107	482	455	460	260	292
4	203 x 203 x 86	893	826	764	513	504
5	203 x 203 x 60	229	220	214	120	127
6	254 x 254 x 73	333	319	320	243	239
7	152 x 152 x 30	120	114	112	71	85
8	305 x 305 x 283	1593	1556	1549	1398	1378

M_c 'Cold' moment capacity (kNm)
Test   Moment capacity calculated using measured temperatures (kNm)
Calc.   Moment capacity calculated using calculated temperatures (kNm)

It can be seen that the calculated and measured temperatures generally compare fairly well with the exception of test 7. The moment capacities are all within about 10% except for test 7 where the error is 20%. It was concluded that for all sections except 152 × 152 the moment capacity calculated using calculated temperatures could be used. A partial safety factor of 1.1 is included to allow for possible unconservatisms in the calculation method. For all 152 × 152 sections a partial factor of 1.2 should be used.

All the fire resistance tests incorporated beams with 15 mm plate. Analyses have been carried out using both 12 mm and 20 mm plate. These have shown that the differences in temperature of the plate and bottom flange are not great when compared to the values calculated using 15 mm plate. For 30 minutes fire resistance all sections will have adequate strength regardless of plate thickness but for 60 minutes there will be a loss of strength if 12 mm plates are used. At present the use of 12 mm plate without fire protection is not recommended.

A theoretical analysis of the transverse bending combined with the longitudinal bending of the plate was carried out. It was concluded that provided the plate (and weld) were satisfactory for normal design then they would also be satisfactory in fire.

5.2 Required strength in fire

The reduced strength of a section in fire is best described using the concept of load ratio defined as:

$$\text{Load ratio, } R = \frac{\text{Moment at fire limit state}}{\text{Moment capacity at 20°C}}$$

For normal design, partial safety factors for loads of 1.4 for dead loads and 1.6 for imposed loads are used. In fire, BS 5950: Part 8 [11] states that these partial factors can both be reduced to 1.0 and any imposed load of a non-permanent nature can have a factor of 0.8. Making the assumption that dead and imposed loads are equal leads to the following limits of load ratio.

Assuming all imposed load is non-permanent $\quad R = \dfrac{1.8}{1.4 + 1.6} = 0.60$

Assuming all imposed load is permanent $\quad R = \dfrac{2}{1.4 + 1.6} = 0.67$

The maximum possible load ratio is 0.71 and corresponds to a beam carrying dead loads only.

In practice, although bending behaviour is critical in fire, it is often not the governing mode for normal design. It is rare therefore for a practical beam to have a load ratio exceeding 0.6 in fire and values in the range 0.5 to 0.55 are most common.

5.3 Composite beams

Composite slim floor beams, Type 3, do not perform as well in fire as non-composite slim floor beams. In order to achieve a fire resistance of 60 minutes a composite beam can only carry proportionately about ⅔ of the load that a non-composite beam can carry. This is because for normal design, a non-composite slim floor beam is very asymmetric and not particularly efficient with the plastic neutral axis low in the section. In fire, as the bottom plate is heated and loses strength, the section effectively becomes more symmetric and more efficient. However a composite section for normal design is effectively a more symmetric section because of the compression in the slab. In fire, as the plate loses strength, it tends to become asymmetric and less efficient. The loss of strength is illustrated in Table 9. In the table the moment capacities for normal design and fire for a 254 × 254 × 73 UC are compared. The temperature distribution across the section is assumed to be at 60 minutes and is the same for both designs. In practice, temperatures in a composite beam will be slightly lower than in a non-composite beam but further tests would be required to evaluate this.

Table 9 *Comparison of composite and non-composite performance*

Design Type	Moment Capacity (kNm)		Load Ratio
	Normal	Fire	
Composite	930	389*	0.42
Non-composite	333	218*	0.65

*Including partial factor of 1.1

It can be seen that although the capacity of the composite beam in fire is almost twice that of the non-composite beam the available load ratio at 60 minutes fire resistance is appreciably less. Because of this poorer performance in fire, designers may have to over-design composite slim floor beams for the normal design situation if they are to be useful without fire protection. The situation is however slightly better for beams designed with a low degree of shear connection.

5.4 Design program

The design program for slim floors includes a check on the load carrying capacity in fire. For 60 minutes fire resistance the moment capacity of the section is calculated and compared to the required capacity at the fire limit state.

The strength of the cross-section in fire is calculated using the moment capacity method. For composite beams the degree of shear connection is taken into account. For each section a full thermal analysis has been carried out and the resulting temperatures have been incorporated into the design program. The temperatures are based on a half encased section and are slightly conservative for the fully composite Type 3 design.

5.5 Recommendations

The recommendations for fire resistance of SLIMFLOR beams are summarised in Table 10.

Table 10 *Summary of recommendations*

Fire Resistance (mins)	Design Type		
	1	2	3
30	No fire check necessary		
60	Moment capacity from Table 11 or use design program		Design program
> 60	Fire protect bottom plate		

5.5.1 30 minutes fire resistance

Any design will achieve 30 minutes fire resistance provided the bottom plate thickness is not less than 12 mm.

5.5.2 60 minutes fire resistance

(a) Types 1 and 2

For sections using plates not less than 15 mm thick, the moment capacity at the fire limit state may be taken from Table 11 or alternatively the design program may be used.

(b) Type 3

It is not feasible to produce a simple design table for composite sections. It is therefore recommended that the design program is used.

5.5.3 90 minutes fire resistance

For more than 60 minutes fire resistance it is necessary to fire protect the bottom plate and specialists in fire protection should be consulted.

5.5.4 Welds

In all cases 8mm fillet welds are recommended.

5.5.4 Edge beams

For edge beams in which both sides of the universal column or channel section are encased in concrete or grout which extends to half the depth of the section then the above recommendations apply. If there is not such an encasement an additional partial safety factor of loads of 1.1 should be applied.

Table 11 *Moment capacity and maximum load ratio for Types 1 and 2 to achieve 60 minutes fire resistance*

Section	Moment Capacity (kNm)		Maximum Load Ratio
	Grade 43	Grade 50	
152 x 152 x 30	55	70	0.59
152 x 152 x 37	67	87	0.59
203 x 203 x 46	113	sc	0.65
203 x 203 x 52	129	166	0.65
203 x 203 x 60	149	193	0.65
203 x 203 x 71	178	232	0.67
203 x 203 x 86	222	289	0.68
254 x 254 x 73	218	sc	0.65
254 x 254 x 89	265	345	0.67
254 x 254 x 107	329	427	0.68
254 x 254 x 132	427	555	0.71
254 x 254 x 167	573	746	0.74
305 x 305 x 118	419	545	0.67
305 x 305 x 137	502	653	0.69
305 x 305 x 158	598	778	0.71
305 x 305 x 198	802	1044	0.74
305 x 305 x 240	1028	1338	0.77
305 x 305 x 283	1207	1607	0.79
356 x 368 x 153	635	sc	0.68
356 x 368 x 177	754	981	0.70
356 x 368 x 202	889	1156	0.72
356 x 406 x 235	1070	1392	0.74
356 x 406 x 287	1383	1799	0.77
356 x 406 x 340	1647	2193	0.79
356 x 406 x 393	1947	2592	0.80
356 x 406 x 467	2383	3174	0.81
356 x 406 x 551	2778	3682	0.82
356 x 406 x 634	3276	4342	0.83

sc: Semi-compact section

All sections which are semi-compact for both steel grades have been omitted. Moment capacities include partial factor of 1.1 or 1.2, as appropriate.

6. DESIGN CHARTS FOR INITIAL SIZING

This section contains 6 Tables where each Table gives Universal Column sizes in grades 43 or 50 for the three types of "SLIMFLOR" construction. The Tables only cover internal beams and are to be used for the purposes of initial sizing. Where final designs are required the beam size should be verified by hand calculation or computer program. In all cases the beam size shown in the Tables has been based on the assumption that an internal bay can be fully erected without the necessity for adjacent bays to provide a balanced load condition in the construction stage. In some cases further economies in beam selection can be made where partial erection of the pc units in the bay can be tolerated. This will largely depend on the site conditions and strict erection procedures (see Section 4).

Types 1 and 2 Tables are self explanatory and should cover the majority of grid sizes encountered in practice. The Tables for Type 3 construction (composite) depend mainly on the depths of in situ concrete, pc units and beam. The amount of possible variation with this form of construction makes it very difficult to produce concise Tables unless all of the constraints are previously known. Therefore the section sizes shown in Tables 16 and 17 are inserted for guidance purposes only. A reasonable guide to the number of shear stud connectors, is 1½ to 2 per square metre of floor area.

The design criteria are listed in Appendix C.

Table 12

TYPE 1	UNIVERSAL COLUMN SIZES IN : GRADE 43					
	IMPOSED LOADING = 3.5 kN/m²					
Beam Span m	PRECAST UNIT SPAN, m					
	4.5		6.0		7.5	
4.5	203 x 203 x 46	h	203 x 203 x 60	h	203 x 203 x 86	h
6.0	203 x 203 x 71	h	254 x 254 x 107	h	254 x 254 x 132	h
7.5	254 x 254 x 107	h	254 x 254 x 132	h	254 x 254 x 167	h
pc unit depth mm	150		150		200	
self wt of unit kN/m²	2.2		2.2		2.7	

Notes:

(1) Key to above Table:

Section Size	Design criterion

    (See Appendix C for design criteria)

(2) The above Table should be read in conjunction with the Tables shown in Section 5 for fire resistance.

(3) The spans and self weights of the pc units have been based on manufacturers literature.

(4) For other loadings see Worked Example No. 1 in Appendix A.

(5) Assumed thickness of bottom flange plate = 15 mm. Plate width based on (B + 200) and rounded-up to the nearest 10mm.

Table 13

TYPE 1	UNIVERSAL COLUMN SIZES IN : GRADE 50					
	IMPOSED LOADING = 3.5 kN/m²					
Beam Span m	PRECAST UNIT SPAN, m					
	4.5		6.0		7.5	
4.5	152 x 152 x 37	d	203 x 203 x 52	h	203 x 203 x 71	h
6.0	203 x 203 x 60	h	203 x 203 x 71	h	254 x 254 x 107	h
7.5	254 x 254 x 89	h	254 x 254 x 107	h	254 x 254 x 132	k
pc unit depth mm	150		150		200	
self wt of unit kN/m²	2.2		2.2		2.7	

Notes:

(1) Key to above Table:

Section Size	Design criterion

  (See Appendix C for design criteria)

(2) The above Table should be read in conjunction with the tables shown in Section 5 for fire resistance.

(3) The spans and self weights of the pc units have been based on manufacturers literature.

(4) For other loadings see Worked Example No. 1 in Appendix A.

(5) Assumed thickness of bottom flange plate = 15 mm. Plate width based on (B + 200) and rounded-up to the nearest 10mm.

Table 14

TYPE 2	UNIVERSAL COLUMN SIZES IN : GRADE 43							
	IMPOSED LOADING = 4.5 kN/m²							
Beam Span m	PRECAST UNIT SPAN, m							
	4.5		6.0		7.5		9.0	
4.5	203 x 203 x 46	d	203 x 203 x 60	d	203 x 203 x 71	d	254 x 254 x 89	d
6.0	203 x 203 x 60	d	203 x 203 x 86	d	254 x 254 x 107	d	254 x 254 x 167	d
7.5	254 x 254 x 73	d	254 x 254 x 107	d	254 x 254 x 167	d	305 x 305 x 198	d
9.0	254 x 254 x 132	k	254 x 254 x 167	k	305 x 305 x 198	k	305 x 305 x 283	k
pc unit depth mm	150		150		200		260	
self wt of unit kN/m²	2.2		2.2		2.7		4.0	

Notes:

(1) Key to above Table:

Section Size	Design criterion

(2) The above Table should be read in conjunction with the tables shown in Section 5 for fire resistance.

(3) The spans and self weights of the pc units have been based on manufacturers literature.

(4) Service and finishes load = 0.25 kN/m². For other loadings see Worked Example No. 2 in Appendix A.

(5) Assumed thickness of bottom flange plate = 15 mm. Plate width based on (B + 200) and rounded-up to the nearest 10mm.

(See Appendix C for design criteria)

Table 15

TYPE 2	UNIVERSAL COLUMN SIZES IN : GRADE 50							
	IMPOSED LOADING = 4.5 kN/m²							
Beam Span m	PRECAST UNIT SPAN, m							
	4.5		6.0		7.5		9.0	
4.5	152 x 152 x 37	d	203 x 203 x 52	d	203 x 203 x 60	d	203 x 203 x 86	d
6.0	203 x 203 x 52	g	203 x 203 x 71	d	254 x 254 x 89	d	254 x 254 x 132	d
7.5	203 x 203 x 86	k	254 x 254 x 89	g	254 x 254 x 132	d	305 x 305 x 158	d
9.0	254 x 254 x 132	k	254 x 254 x 167	k	305 x 305 x 158	k	305 x 305 x 283	k
pc unit depth mm	150		150		200		260	
self wt of unit kNm²	2.2		2.2		2.7		4.0	

Notes:

(1) Key to above Table:

Section Size	Design criterion

(See Appendix C for design criteria)

(2) The above Table should be read in conjunction with the tables shown in Section 5 for fire resistance.

(3) The spans and self weights of the pc units have been based on manufacturers literature.

(4) Services and finishes load = 0.25 kN/m². For other loadings see Worked Example No. 2 in Appendix A.

(5) Assumed thickness of bottom flange plate = 15 mm. Plate width based on (B + 200) and rounded-up to the nearest 10 mm.

Table 16

TYPE 3	UNIVERSAL COLUMN SIZES IN : GRADE 43
	IMPOSED LOADING = 6.0 kN/m²

Beam Span m	PRECAST UNIT SPAN, m								
	6.0			7.5			9.0		
4.5	95	203 × 203 × 52	d	85	254 × 254 × 73	d	90	254 × 254 × 89	d
	2.7	200		4.0	260		4.0	260	
6.0	140	254 × 254 × 73	d	105	254 × 254 × 132	d	110	305 × 305 × 137	d
	2.7	200		4.0	260		4.0	300	
7.5	165	254 × 254 × 132	g	115	305 × 305 × 158	d			
	2.7	200		4.0	300				
9.0	125	305 × 305 × 198	d						
	4.0	300							
10.0	140	305 × 305 × 240	k						
	4.0	300							

Notes:

(1) Key to the values shown in the above Table

depth of in situ concrete mm	Section Size	design criterion
weight allowed for pc unit (kN/m²)	pc unit depth	

(See Appendix C for design criteria)

(2) Normal weight, grade 30 concrete has been assumed.
(3) The above Table should be read in conjunction with the tables shown in Section 5 for fire resistance.
(4) The spans and self weights of the pc units have been based on manufacturers' literature.
(5) Services and finishes load = 0.25 kN/m². For other loadings see Worked Example No. 3.

Table 17

TYPE 3	UNIVERSAL COLUMN SIZES IN GRADE 50
	IMPOSED LOADING = 6.0 kN/m²

Beam Span m	PRECAST UNIT SPAN, m											
	6.0			7.5			9.0			10.0		
4.5	95	203 x 203 x 52	d	90	254 x 254 x 89	w	90	254 x 254 x 89	d	100	305 x 305 x 118	v
	2.7	200		4.0	260		4.0	260		4.0	300	
6.0	105	203 x 203 x 71	d	95	254 x 254 x 107	d	100	305 x 305 x 118	d	100	305 x 305 x 118	d
	2.7	200		4.0	260		4.0	300		4.0	300	
7.5	155	254 x 254 x 107	c	115	254 x 254 x 167	d	115	305 x 305 x 158	d	125	305 x 305 x 198	d
	2.7	200		4.0	260		4.0	300		4.0	300	
9.0	155	305 x 305 x 158	c	125	305 x 305 x 198	k	140	305 x 305 x 240	k			
	4.0	260		4.0	300		4.0	300				
10.0	140	305 x 305 x 240	k	180	305 x 305 x 283	k						
	4.0	300		4.0	300							

Notes:

(1) Key to the values shown in the above Table

depth of in situ concrete mm	Section Size	design criterion
weight allowed for pc unit (kN/m²)	pc unit depth	

(2) Normal weight, grade 30 concrete has been assumed.
(3) The above Table should be read in conjunction with the tables shown in Section 5 for fire resistance.
(4) The spans and self weights of the pc units have been based on manufacturers' literature.
(5) Services and finishes load = 0.25 kN/m². For other loadings see Worked Example No. 3.

(See Appendix C for design criteria).

REFERENCES

1. BRITISH STANDARDS INSTITUTION
   BS 5950: Structural use of steelwork in building
   Part 1: Code of practice for design in simple and continuous construction: hot rolled sections
   BSI, 1990

2. BRITISH STANDARDS INSTITUTION
   BS 5950: Structural use of steelwork in building
   Part 3: Codes of practice for design in composite construction
   Section 3.1: Design of simple and continuous composite beams
   BSI, 1990

3. BRETT, P.A. and RUSHTON, J.A.
   Parallel beam approach - a design guide
   The Steel Construction Institute, 1990

4. DEPARTMENT OF THE ENVIRONMENT
   The Building Regulations 1985
   Approved Documents to A3 - Disproportionate collapse
   HMSO, 1985

5. BAXTER BROWN, J. M. C. D.
   Introductory Solid Mechanics
   John Wiley & Sons, 1973

6. NETHERCOT, D.A., SALTER, P.R. and MALIK, A.S.
   Design of members subject to combined bending and torsion
   The Steel Construction Institute, 1989

7. LAWSON, R.M.
   Design of composite slabs and beams with steel decking
   The Steel Construction Institute, 1989

8. LAWSON, R. M.
   Commentary to BS 5950: Part 3: Section 3.1 "Composite beams"
   The Steel Construction Institute, 1990

9. WYATT, T.A.
   Design guide on the vibration of floors
   The Steel Construction Institute, 1989

10. THE STEEL CONSTRUCTION INSTITUTE
    Joints in simple construction, Volume 1: design methods
    SCI/BCSA, 1991

11. BRITISH STANDARDS INSTITUTION
    BS 5950: The structural use of steel in building
    Part 8: Code of practice for fire resistant design
    BSI, 1990

12. BRITISH STANDARDS INSTITUTION
    BS 6399: 1984 Design loading for buildings
    Part 1: Code of practice for dead and imposed loads
    BSI, 1984

13. THE STEEL CONSTRUCTION INSTITUTE
    Steelwork design guide to BS 5950: Part 1: 1985
    Volume 1: Section properties and member capacities. 2nd Edition
    SCI, 1987

APPENDIX A: Worked examples

The purpose of the worked examples is to illustrate the design procedures for the three methods of construction using the "SLIMFLOR" beam.

The three designs will be based on the same Universal Column section and flange plate. This is to reduce the amount of calculation for section properties etc.

The UC beam is not necessarily the absolute minimum section required for the given parameters. With additional refinement (using design charts and computer program) it may be possible to show that the UC beam used for the three examples is larger than required.

Beam size: 305 × 305 × 118 UC   Grade 50
Flange size: 510 × 15 thick   Grade 50

Worked example	Grid size (internal bay)	Imposed Loading kN/m^2 (Occupancy & Partition)	Fire Resistance
1.	7.5 m × 7.5 m	2.5 + 1	30 mins
2.	8.0 m × 8.0 m	3.5 + 1	60 mins
3.	8.0 m × 8.0 m	5 + 1	60 mins

General Assumptions

(1) For all three types of construction, no restraint to the compression flange throughout the length of the beam has been assumed in the construction stage.

(2) In the construction stage the beam is assumed to be torsionally fixed and warping free at the supports.

(3) For the imposed loading condition the beam may be assumed to be laterally restrained, provided the void around the UC is filled to at least half the beam depth with grout or similar material. This applies to Type 1 construction, Types 2 and 3 use in situ concrete.

(4) The combination of bending and torsion has to be considered for the imposed loading (out of balance) condition. The beam has been assumed to be laterally restrained for the three forms of construction, see (3) above. The torsion which can arise from the out of balance imposed loads has been assumed to have little effect on the floor structure for Type 2 form of construction. The partial filling of the void that surrounds the beam for Type 1 form of construction is considered to offer no resistance to these torsional actions. The design check for this case where the beam is laterally restrained but subject to torsion is known as the "capacity" check. Unfortunately, no simplified version can be given for the capacity check and the full rigorous analysis will have to be used. These procedures are given in Appendix A for the first worked example.

TYPE 1 CONSTRUCTION

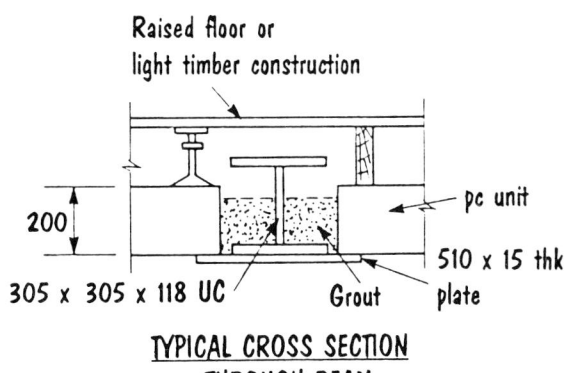

TYPICAL CROSS SECTION
THROUGH BEAM

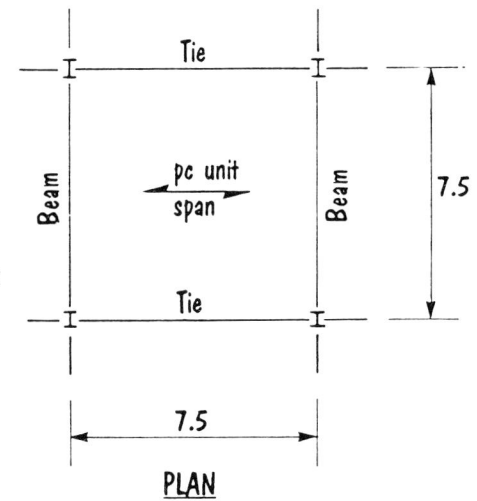

PLAN
INTERNAL BAY

LOADING - DEAD	kN/m²
PC units	2.67
Self wt of steel	0.20
Construction load	0.50
Floor & minor services	0.25
Grouting to beam (to half depth)	0.15

LOADING - IMPOSED	
Occupancy	2.5
Partitions	1.0
	3.5

Imposed load reduction to BS 6399: Part 1 (See Reference 12)

Supported Area = 7.5 × 7.5
= 56.25 m²

Reduction
= 3.5 × 0.056 = 0.2 kN/m²

∴ Revised imposed load
= 3.5 - 0.2
= 3.3 kN/m²

STEEL STRENGTH

For flange thickness of 18.7 mm, design strength of grade 50 steel, $p_y = 345$ N/mm² (refer to BS5950: Part 1)

	Job No.	PUB 810	Sheet	2 of 56	Rev.
The Steel Construction Institute	Job Title	Worked Example No 1			
Silwood Park Ascot Berks SL5 7QN Telephone: (0344) 23345 Fax: (0344) 22944	Subject	Load Combinations			
CALCULATION SHEET	Client		Made by DLM	Date Oct 1991	
			Checked by JWR	Date Oct 1991	

LOAD COMBINATIONS
Design for:

CASE 1

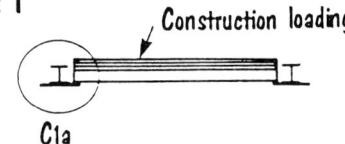

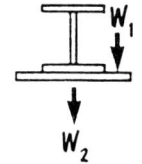

a) Lateral Torsional Buckling check (LTB) has to be combined with torsion. See Section 4 of this publication. It may be necessary to limit the number of pc units placed on one side of the beams.

C1a

CASE 2

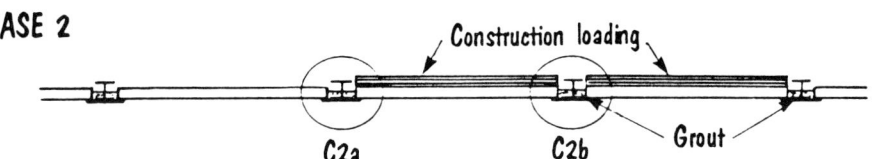

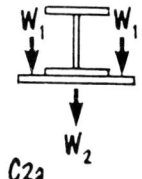

C2a

a) LTB has to be combined with torsion for construction out of balance loading.

b) Max. loading condition for LTB during construction. Section is grouted at this stage.

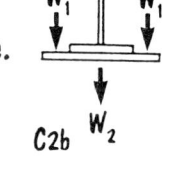

CASE 3

C2b

a) Beam is considered to be laterally restrained (grouted in accordance with section 3 of this publication). Torsion has to be combined with bending using the 'capacity check' (see Note 4 - Assumptions).

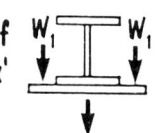

b) Beam is considered to be laterally restrained. Max. factored moment is applied to beam.

C3a

NB:

1. Where LTB has to be combined with torsion this will govern when compared to the 'capacity check'.

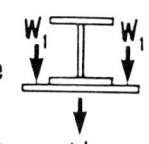

2. When the member is considered as laterally restrained then the 'capacity check' will govern.

C3b

Note: The imposed condition shown in Case 3a only applies to Type 1 construction. Types 2 and 3 construction are capable of transferring the out of balance forces across the section.

The Steel Construction Institute 	Job No. PUB 810		Sheet 3 of 56	Rev.
Silwood Park Ascot Berks SL5 7QN Telephone: (0344) 23345 Fax: (0344) 22944	Job Title	Worked Example No 1		
	Subject	Load Combinations cont'd		
CALCULATION SHEET	Client	Made by DLM	Date Oct 1991	
		Checked by JWR	Date Oct 1991	

CASE 1 - CONSTRUCTION LOADING

(C1a) $W_1 = [(2.67 \times 1.4) + (0.5 \times 1.6)] \, 7.5^2/2$
$= 127.6$ kN
$W_2 = 0.2 \times 1.4 \times 7.5^2$
$= 15.8$ kN

Checks: LTB combined with torsion

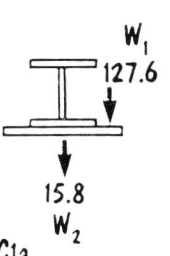

CASE 2 - CONSTRUCTION LOADING

(C2a) $W_1 = 0.5 \times 1.6 \times 7.5^2/2$
$= 22.5$ kN
$W_2 = (2.67 + 0.2 + 0.15) \, 7.5^2 \times 1.4$
$= 237.8$ kN

Checks: LTB combined with torsion

(C2b) $W_2 = [(0.5 \times 1.6) + (2.67 + 0.2 + 0.15)1.4] \, 7.5^2$
$= 282.8$ kN

Checks: Max. loading condition for LTB

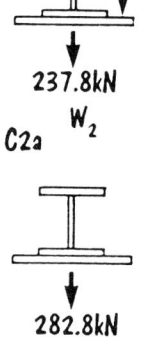

CASE 3 - IMPOSED LOADING

(C3a) $W_1 = 3.3 \times 1.6 \times 7.5^2/2$
$= 148.5$ kN
$W_2 = [(2.67 + 0.2 + 0.25 + 0.15)1.4] \, 7.5^2$
$= 257.5$ kN

Checks: Beam restrained, combine with torsion use 'capacity check'

(C3b) $W_2 = [(3.3 \times 1.6) + (2.67 + 0.2 + 0.25 + 0.15)1.4] \, 7.5^2$
$= 554.5$ kN

Checks: Max. factored moment. Use 'capacity check'.

By inspection, it is not possible to assume which design criterion governs. Therefore, for the purposes of this design example and to illustrate the procedures, the design will be carried out in full.

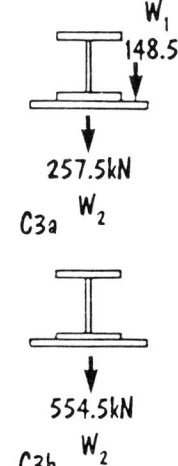

The Steel Construction Institute		Job No.	PUB 810	Sheet	4 of 56	Rev.
Silwood Park Ascot Berks SL5 7QN Telephone: (0344) 23345 Fax: (0344) 22944		Job Title	Worked Example No 1			
		Subject	Lateral Torsional Buckling			
CALCULATION SHEET		Client		Made by DLM		Date Oct 1991
				Checked by JWR		Date Oct 1991

LATERAL TORSIONAL BUCKLING

$M_b = S_x p_b$

Determine, p_b (Ref. BS 5950: Part 1)

firstly, λ_{LT} has to be evaluated: $\lambda_{LT} = nuv\lambda$

where: n is taken as 1.0

$$u = \left(\frac{4S_x^2 \, Y}{A^2 \, h_s^2}\right)^{\frac{1}{4}}$$

$$v = \left[(4N(1-N) + \frac{1}{20}\left(\frac{\lambda}{x}\right)^2 + \psi^2)^{\frac{1}{2}} + \psi\right]^{-\frac{1}{2}}$$

$\lambda = L_E/r_y$

SECTION PROPERTIES

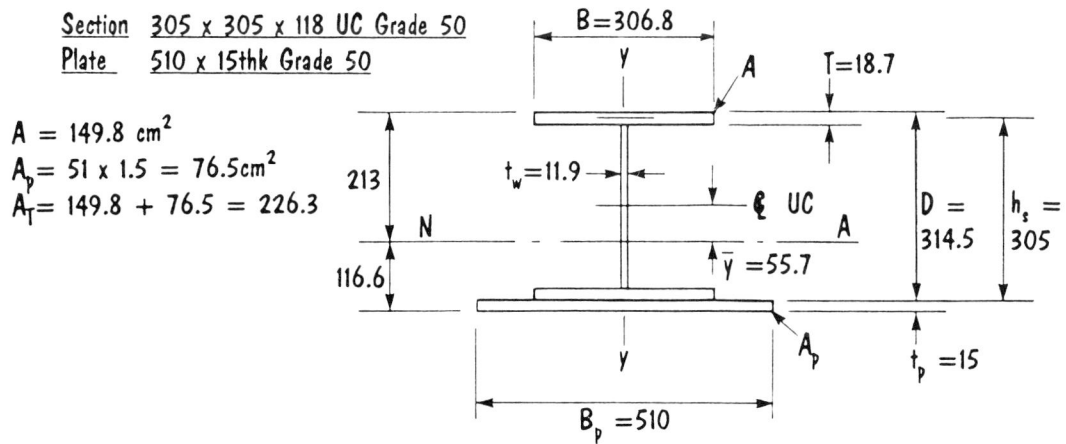

Section 305 x 305 x 118 UC Grade 50
Plate   510 x 15thk Grade 50

$A = 149.8 \text{ cm}^2$
$A_p = 51 \times 1.5 = 76.5 \text{ cm}^2$
$A_T = 149.8 + 76.5 = 226.3$

Total area $= 149.8 + 76.5 = 226.3 \text{ cm}^2$

$$\bar{y} = \frac{(314.5 + 15)76.5}{2 \times (149.8 + 76.5)} = 55.7 \text{ mm}$$

The Steel Construction Institute	Job No.	PUB 810		Sheet 5 of 56	Rev.
Silwood Park Ascot Berks SL5 7QN Telephone: (0344) 23345 Fax: (0344) 22944	Job Title	Worked Example No 1			
	Subject	Lateral Torsional Buckling cont'd			
CALCULATION SHEET	Client		Made by DLM	Date	Oct 1991
			Checked by JWR	Date	Oct 1991

Second moment of area of plated section in major axis direction.

$$I_{xx} = (27601 + 149.8 \times 5.57^2) + 76.5 \left(\frac{314.5}{20} + \frac{15}{20} - \frac{55.7}{10}\right)^2$$
$$= 41347 \text{ cm}^4$$

$$I_{yy} = 9006 + \frac{1.5 \times 51^3}{12}, \quad r_{yy} = \left(\frac{25587}{226.3}\right)^{\frac{1}{2}} = 10.63 \text{ cm}$$
$$= 25587 \text{ cm}^4$$

Second moment of area of flanges in transverse bending

$$I_{cf} = TB^3/12 = 1.87 \times 30.68^3/12 = 4500.1 \text{ cm}^4$$

$$I_{tf} = TB^3/12 + t_pB_p^3/12 = 4500.1 + 1.5 \times 51^3/12 = 21082 \text{ cm}^4.$$

BUCKLING PARAMETER, u

$$u = \left(\frac{4 S_x^2 \gamma}{A^2 h_s}\right)^{\frac{1}{4}}$$

Plastic modulus of the section about the x x axis, S_x. This can be calculated using the moment capacity formula shown in Appendix B.

$R_s = 149.8 \times 345/10 = 5168.1$ kN
$R_p = 76.5 \times 345/10 = 2639.3$ kN

$(R_s - R_p) = 2528.8$ kN

$$y = \frac{2528.8 \times 10^3}{2 \times 306.8 \times 345} = 11.9 \text{ mm}$$

$$R_w = R_s - 2R_f = 5168.1 - 2 \times \frac{306.8}{10^3} \times 18.7 \times 345$$
$$= 1210 \text{ kN}$$

The Steel Construction Institute	Job No.	**PUB 810**		Sheet **6** of **56**	Rev.
Silwood Park Ascot Berks SL5 7QN Telephone: (0344) 23345 Fax: (0344) 22944	Job Title	**Worked Example No 1**			
	Subject	**Lateral Torsional Buckling cont'd**			
CALCULATION SHEET	Client	Made by **DLM**		Date **Oct 1991**	
		Checked by **JWR**		Date **Oct 1991**	

$R_p > R_w$ & $R_s > R_p$

∴ Case 2SP from Appendix B

Note: The exact value for h_s is the distance between the individual centres of gravity of the top and bottom flanges. The neutral axis of the combined thicknesses (T and t_p) is assumed to be at the top surface of the bottom flange plate.

$$M_c = \frac{5168.1}{10^3} \times \frac{314.5}{2} + \frac{2639.3}{10^3} \times \frac{15}{2} - \frac{2528.8^2}{4 \times 306.8 \times 345}$$

$$812.7 \quad + 19.8 \quad - 15.1$$

$$= 817.4 \text{ kN.m}$$

$$\therefore S_x = \frac{817.4 \times 10^6}{345 \times 10^3} = 2369.3 \text{ cm}^3$$

Note: As an alternative the section modulus can be obtained from first principles.

e.g. $(76.5 \times 1.94) + (1.19 \times 30.68) 1.19/2 + (35.06 \times 14.535)$
$+ (0.68 \times 30.68) 0.34 + (30.68 \times 1.87) 29.33 = 2369.5 \text{ cm}^3$

$$\gamma = 1 - I_y/I_x = 1 - \frac{25587}{41347} = 0.381$$

$$A_T = 149.8 + 76.5 = 226.3 \text{ cm}^2$$

$$\therefore u = \left(\frac{4 \times 2369.3^2 \times 0.381}{226.3^2 \times 30.5^2} \right)^{\frac{1}{4}} = 0.652$$

Buckling parameter, $u = 0.652$

SLENDERNESS FACTOR, v

$$v = [(4N(1-N) + 1/20 \, (\lambda/x)^2 + \psi^2)^{\frac{1}{2}} + \psi]^{-\frac{1}{2}}$$

	Job No.	PUB 810		Sheet	7	of	56	Rev.
The Steel Construction Institute Silwood Park Ascot Berks SL5 7QN Telephone: (0344) 23345 Fax: (0344) 22944	Job Title	Worked Example No 1						
	Subject	Lateral Torsional Buckling cont'd						
CALCULATION SHEET	Client		Made by DLM			Date Oct 1991		
			Checked by JWR			Date Oct 1991		

where: $N = \dfrac{I_{cf}}{I_{cf} + I_{tf}} = \dfrac{4500.1}{25583} = 0.176$

$N < 0.5$, $\psi = 1.0\,(2N-1) = 2 \times 0.176 - 1 = -0.648$

TORSIONAL INDEX, x

$x = 0.566\, h_s\, (A/J)^{1/2}$

Torsional constant of cross-section, J

$J = J_{uc} + \tfrac{1}{3}(t_p^3 . B_p)$

$= 160 + \tfrac{1}{3}(1.5^3 \times 51) = 217.3 \text{ cm}^4$

$\therefore x = 0.566 \times 30.5 \left(\dfrac{226.3}{217.3}\right)^{\tfrac{1}{2}} = 17.61$

$\lambda = L_E/r_y \quad L_E = 1.0 \quad$ (Note: loads applied through the tension flange are 'stabilizing' loads, but this benefit is not included)

$= \dfrac{7.5 \times 10^3 \times 1.0}{106.3}$

$= 70.6$

$\lambda/x = 70.6/17.61 = 4.0$

Using Table 14 $v = 1.15$

Check using formulae for v

$v = \left[\left(4 \times 0.176\,(1 - 0.176) + \dfrac{1}{20} \times 4^2 + 0.648^2\right)^{1/2} - 0.648\right]^{-1/2}$

$= 1.2$ (Checks out with BS5950: Part 1, Table 14)

	Job No.	PUB 810		Sheet 8 of 56	Rev.
The Steel Construction Institute	Job Title	Worked Example No 1			
Silwood Park Ascot Berks SL5 7QN Telephone: (0344) 23345 Fax: (0344) 22944	Subject	Lateral Torsional Buckling cont'd			
	Client		Made by DLM		Date Oct 1991
CALCULATION SHEET			Checked by JWR		Date Oct 1991

∴ λ_{LT} = 1.0 × 0.651 × 1.2 × 70.6

= 55.2

From BS5950: Part 1, Table 11 p_b = 268 N/mm²

∴ M_b = $\dfrac{2369.3 \times 10^3 \times 268}{10^6}$

= 635.0 kN.m

Before proceeding with further design checks determine the biaxial stress effects (if any) in the plate. In some instances the transverse moment may have such an effect as to reduce the moment capacity for overall bending of the section.

Design Case - Imposed Loading
(C3b)

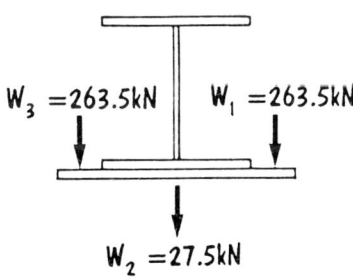

W_1 = W_3 = [(2.67 + 0.25)1.4 + (3.3 × 1.6)]7.5²/2 = 263.5 kN

W_2 = (0.2 + 0.15)1.4 + 7.5² = 27.5 kN

Total factored load = 2 × 263.5 + 27.5 = 554.5 kN

M_x = 554.5 × 7.5/8 = 519.8 kN.m

Maximum transverse moment in plate is obtained when:

$\dfrac{M}{M_p} = \dfrac{c^2 - \sigma_1^2}{2c\, p_y}$ where: $c = (4p_y^2 - 3\sigma_1^2)^{1/2}$

Commentary to calculation sheet

To simplify the transverse bending effects, assume the moment diagram across the bottom flange of the UC is constant.

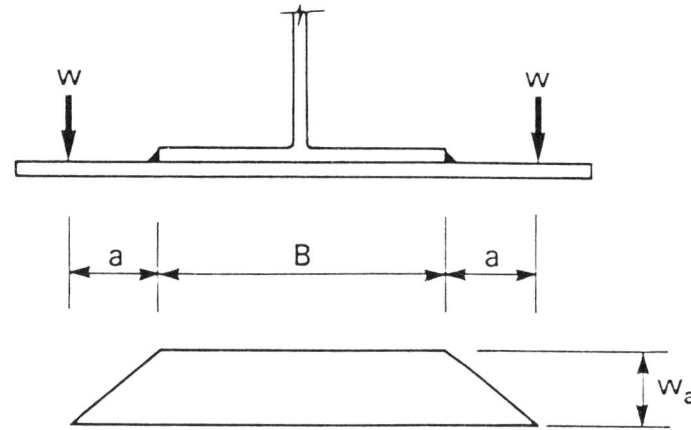

The Steel Construction Institute 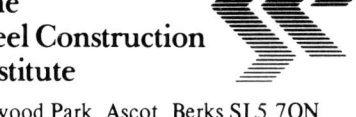 Silwood Park Ascot Berks SL5 7QN Telephone: (0344) 23345 Fax: (0344) 22944 **CALCULATION SHEET**	Job No. **PUB 810**		Sheet **9** of **56**	Rev.
	Job Title	**Worked Example No 1**		
	Subject	**Biaxial Stress Effects**		
	Client	Made by **DLM**		Date **Oct 1991**
		Checked by **JWR**		Date **Oct 1991**

Note: The transverse bending effects refer directly to the bottom flange plate.

Hence, 15mm < 16mm use $p_y = 355$ N/mm²

$$\sigma_1 = \frac{M_x}{S_x} = \frac{519.8 \times 10^6}{2369.3 \times 10^3} = 219.4 \text{ N/mm}^2$$

$$c = (4 \times 355^2 - 3 \times 219.4^2)^{1/2} = 599.7 \text{ N/mm}^2$$

$$\frac{M}{M_p} \leq \frac{599.7^2 - 219.4^2}{2 \times 599.7 \times 355} = \text{i.e. } 0.732$$

∴ The ratio of $\frac{\text{plate transverse moment}}{\text{moment capacity of plate}}$ must not exceed the 0.732 otherwise the longitudinal bending capacity of the beam is reduced.

PLATE BENDING

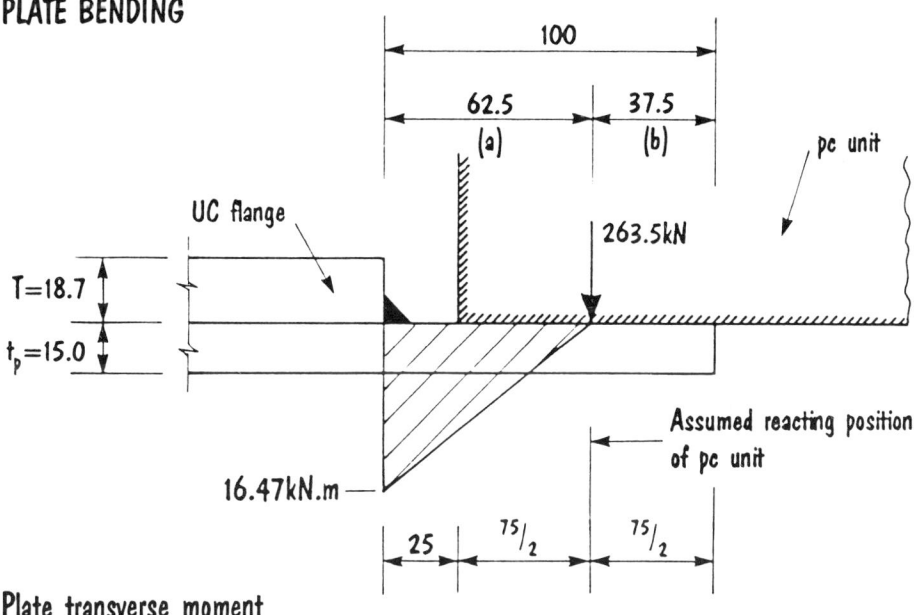

Plate transverse moment

$M = 263.5 \times 62.5/10^3 = 16.47$ kN.m (Total)

for unit length, moment $= 16.47/7.5 = 2.2$ kN.m per m

	Job No.	PUB 810		Sheet	10 of 56	Rev.
The Steel Construction Institute Silwood Park Ascot Berks SL5 7QN Telephone: (0344) 23345 Fax: (0344) 22944	Job Title	Worked Example No 1				
	Subject	Biaxial Stress Effects cont'd Case 1 Construction Stage - Torsion				
CALCULATION SHEET	Client		Made by	DLM	Date	Oct 1991
			Checked by	JWR	Date	Oct 1991

PLATE CAPACITY

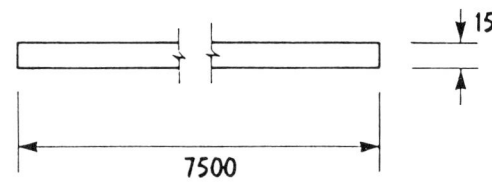

$$S_{xp} = \frac{t_p^2 L_1}{4} = \frac{1.5^2 \times 750}{4} = 421.9 \text{ cm}^3 \text{ (Total)}$$

for unit length = 421.9/7.5 = 56.25 cm³ per m

Plate moment capacity $M_p = \dfrac{56.25 \times 355}{10^3} = 19.97$ kN.m per m

$$\frac{M}{M_p} = \frac{2.2}{19.97} = 0.11 < 0.732 \text{ OK}$$

∴ In this case biaxial bending has no influence on the overall bending of the section.

CASE 1 CONSTRUCTION CONDITION

$\left(\dfrac{306.8}{2}\right) + 62.5 = 216$

W_1 127.6 kN

15.8 kN
W_2

Case C1a

Check using the simple approach

Where $z = L_1 = 7.5$ m (This is the worst case for pc units placed on one side of the beam)

Note: When $a = L_1$ then $M = \dfrac{WL}{8}$

	Job No.	PUB 810		Sheet 11 of 56	Rev.
The Steel Construction Institute Silwood Park Ascot Berks SL5 7QN Telephone: (0344) 23345 Fax: (0344) 22944 **CALCULATION SHEET**	Job Title	Worked Example No 1			
	Subject	Case 1 Construction Stage - Torsion cont'd			
	Client		Made by DLM	Date Oct 1991	
			Checked by JWR	Date Oct 1991	

$$M_1 = \frac{127.6 \times 7.5}{2}\left(1 - \frac{7.5}{15}\right)^2 = 119.6 \text{ kN.m}$$

Self weight moment at mid-span

$$M_2 = \frac{15.8 \times 3.75}{15}\left(7.5 - \frac{7.5}{2}\right) = \underline{14.8 \text{ kN.m}}$$

$$M_x = \underline{134.4 \text{ kN.m}}$$

Equivalent transverse force in flanges to resist torsion

$$F = 127.6 \times 216/305 = 90.4 \text{ kN}$$

Transverse bending moment in top flange

$$M_y = \frac{90.4 \times 7.5}{2}(1 - \tfrac{1}{2})^2 = 84.8 \text{ kN.m}$$

Transverse bending capacity of top flange

$$M_{cy} = \frac{1.87 \times 30.68^2}{4} \times \frac{345}{10^3} = 151.8 \text{ kN.m}$$

Unity Factor (UF) $= \dfrac{134.4}{635.0} + \dfrac{84.8}{151.8} = 0.77 < 1.0$ OK (see Section 4.1.5)

This implies that a whole 'internal bay' could be erected without any undue influence of torsion combined with bending. If UF > 1.0 then the amount of pc units erected in one bay would have to be limited before units were erected in adjacent bays.

	Job No.	PUB 810	Sheet	12 of 56	Rev.
The Steel Construction Institute	Job Title	Worked Example No 1			
Silwood Park Ascot Berks SL5 7QN Telephone: (0344) 23345 Fax: (0344) 22944	Subject	Rigorous Analysis for Construction Stage Torsion			
CALCULATION SHEET	Client	Made by DLM		Date Oct 1991	
		Checked by JWR		Date Oct 1991	

Check Case 1 using the rigorous analysis see (Reference 4)

This check is to demonstrate the procedures given in Reference 4 and to compare the values of UF for the simple and rigorous approach. It is an ALTERNATIVE to the preceding analysis.

DETERMINE, a TORSIONAL BENDING CONSTANT

$$a = \left(\frac{EH}{GJ}\right)^{1/2} \quad E/G \text{ taken as } 2.6$$

WARPING CONSTANT, H

$$h \approx D - T/2$$
$$= \frac{314.5}{10} - \frac{18.7}{20} = 30.5 \text{ cm}$$

$$H = \frac{h^2 \, I_{cf} \, I_{tf}}{I_y} = \frac{30.5^2 \times 4500.1 \times 21082}{25587} = 3.45 \times 10^6 \text{ cm}^6$$

$$a = \left(\frac{2.6 \times 3.45 \times 10^6}{217.3}\right)^{\frac{1}{2}} = 203.1 \text{ cm}$$

$$\therefore L_t/a = \frac{7500}{2031} = 3.7$$

CHECK COMBINED BENDING AND TORSION AT ULS

i) Buckling Check

$$\frac{M_x}{M_b} + \left(\frac{\sigma_{byT} + \sigma_w}{p_y}\right)\left(1 + \frac{0.5 \, M_x}{M_b}\right) \le 1.0$$

The Steel Construction Institute	Job No.	PUB 810		Sheet 13 of 56	Rev.
Silwood Park Ascot Berks SL5 7QN Telephone: (0344) 23345 Fax: (0344) 22944	Job Title	Worked Example No 1			
	Subject	Rigorous Analysis cont'd			
CALCULATION SHEET	Client		Made by DLM	Date Oct 1991	
			Checked by JWR	Date Oct 1991	

DETERMINE ∅ AT ULS THEN, σ_{byT}

T_q = $127.6 \times 216/10^3$ = 27.6 kN.m

L_1/a = 3.7 using graph 1 shown on Calculation Sheet 21

$GJ/T_q a \approx 0.28$

Therefore,

$\emptyset_{ULS}$ = $\dfrac{0.28 \times 27.6 \times 10^6 \times 2031}{79000 \times 217.3 \times 10^4}$ = 0.0914 Rads (5.2°)

M_x = 134.4 kN.m

M_{yT} = 134.4×0.0914 = 12.30 kN.m

σ_{byT} = M_{yT}/Z_y Z_y = I_y/y = $\dfrac{25587}{15.34}$ = 1668 cm³

σ_{byT} = $\dfrac{12.30 \times 10^3}{1668}$ = 7.4 N/mm²

Commentary to calculation sheet

It can be demonstrated that the normalised warping function, W_{no} at the point on the cross-section at which peak stress occurs,

$$= \frac{Bh_1}{2} \quad \text{(Top flange)}$$

For information purposes only, W_{no} becomes $\dfrac{B_p.h_2}{2}$ for the bottom flange plate.

See Reference 6 for further information.

	Job No.	PUB 810	Sheet 14 of 56	Rev.
The Steel Construction Institute Silwood Park Ascot Berks SL5 7QN Telephone: (0344) 23345 Fax: (0344) 22944 **CALCULATION SHEET**	Job Title	Worked Example No 1		
	Subject	Rigorous Analysis cont'd		
	Client	Made by DLM	Date Oct 1991	
		Checked by JWR	Date Oct 1991	

WARPING NORMAL STRESS, σ_w

$\sigma_w = E\, W_{no}\, \theta''$

NORMALISED WARPING FUNCTION, W_{no}

W_{no} (top) taken as $\dfrac{Bh_1}{2}$

Compression flange under consideration

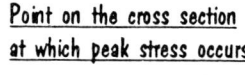

Point on the cross section at which peak stress occurs

Firstly, calculate position of shear centre, Y_o

$Y_o = \dfrac{h_b\, I_{tf} - h_t\, I_{cf}}{I_b + I_t}$ $h_1 = 203.6 + 47.9 = 251.5$ mm

$= \dfrac{10.16 \times 21082 - 20.36 \times 4500.1}{21082 + 4500.1} = 4.79$ cm

$\therefore W_{no}$ (Top) $= \dfrac{306.8 \times 251.5}{2 \times 10^2} = 385.8$ cm^2

Determine, θ'' (using Graph 2 shown on Calculation Sheet 21).

$L_t/a = 3.7$ from graph 2 $0.19 \approx \dfrac{\theta''\, G\, Ja}{Tq}$

$\therefore \theta''_{ULS} = \dfrac{0.19 \times 27.6 \times 10^6}{79000 \times 217.3 \times 10^4 \times 2031} = 1.50 \times 10^{-8}$ Rads

$\therefore \sigma_w = 205000 \times 385.8 \times 1.50 \times 10^{-8} \times 10^2$

$= 119.0$ N/mm^2

	Job No.	PUB 810	Sheet	15 of 56	Rev.
The Steel Construction Institute Silwood Park Ascot Berks SL5 7QN Telephone: (0344) 23345 Fax: (0344) 22944 **CALCULATION SHEET**	Job Title	Worked Example No 1			
	Subject	Rigorous Analysis cont'd			
	Client	Made by DLM		Date Oct 1991	
		Checked by JWR		Date Oct 1991	

BUCKLING CHECK

$$\frac{134.4}{635.0} + \left(\frac{7.4+119.0}{345}\right)\left(1 + \frac{0.5 \times 134.4}{635.0}\right)$$

$0.212 + 0.405 = 0.617 < 1.0$ OK

∴ <u>buckling check satisfactory - Case 1</u>

LOCAL CAPACITY CHECK

$\sigma_{bx} + \sigma_{byT} + \sigma_w < p_y$

$\sigma_{bx} = M_x/Z_x \quad Z_x = I_x/y = \frac{41347}{21.3} = 1941.2 \text{ cm}^3$

$= \frac{134.4 \times 10^3}{1941.2} = 69.2 \text{ N/mm}^2$

Local capacity check

$69.2 + 7.4 + 119.0 = 195.6 \text{ N/mm}^2 < p_y = 345 \text{ N/mm}^2$ (UF = 0.567)

Buckling check governs i.e. $0.617 > 0.567$

This governing value of 0.617 should be compared with the value of 0.77 obtained from the simplified approach. This shows that the simplified approach is conservative and can be used for general design.

<u>Case 1 - satisfactory</u>

Note: When lateral torsional buckling can occur, the 'capacity check' is not critical for uniformly distributed loading.

	Job No.	PUB 810		Sheet 16 of 56	Rev.
The Steel Construction Institute Silwood Park Ascot Berks SL5 7QN Telephone: (0344) 23345 Fax: (0344) 22944 **CALCULATION SHEET**	Job Title	Worked Example No 1			
	Subject	Case 2 Construction Stage - Torsion			
	Client		Made by DLM		Date Oct 1991
			Checked by JWR		Date Oct 1991

CASE 2 - CONSTRUCTION STAGE LOADING

Case 2a

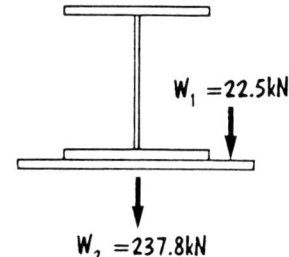

Check using simple approach

M_1 = 22.5 × 7.5/8 = 21.1 kN.m
M_2 = 237.8 × 7.5/8 = 223.0 kN.m
M_x = 244.1 kN.m

Equivalent transverse force in flanges to resist torsion

F = 22.5 × 216/305 = 15.9 kN

M_y = 15.9 × 7.5/8 = 14.9 kN.m

UF = $\dfrac{244.1}{635} + \dfrac{14.9}{151.8}$ = 0.48 < 1.0 OK

<u>Case 2a - satisfactory</u>

Case 2b

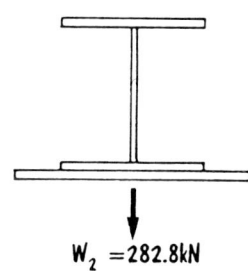

Moment for LTB

M_{ax} = 282.8 × 7.5/8 = 265.1 kN.m < 635.0 kN.m OK

<u>Case 2b - satisfactory</u>

∴ <u>Case 2 - satisfactory</u>

		Job No.	PUB 810		Sheet	17	of	56	Rev.
The Steel Construction Institute 		Job Title	Worked Example No 1						
Silwood Park Ascot Berks SL5 7QN Telephone: (0344) 23345 Fax: (0344) 22944		Subject	Rigorous Analysis for Case 3 Imposed Loading - Torsion						
CALCULATION SHEET		Client		Made by	DLM		Date	Oct 1991	
				Checked by	JWR		Date	Oct 1991	

CASE 3 - IMPOSED LOADING

Case 3a

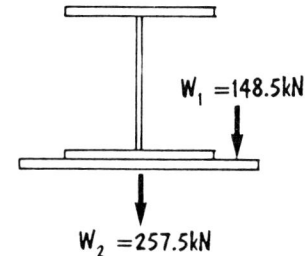

Local capacity check (refer to SCI publication 'Design of members subject to combined bending and torsion')

$\sigma_{bx} + \sigma_{byT} + \sigma_w < p_y$

$\sigma_{bx} = \dfrac{M_x}{Z_x}$, where $M_x = (148.5 + 257.5) \, 7.5/8 = 380.6$ kN.m

$\sigma_{bx} = \dfrac{380.6 \times 10^3}{1941.2} = 196$ N/mm²

$\sigma_{byT} = M_{yT}/Z_y$

Torsional moment

$T_q = 148.5 \times 216/10^3 = 32.10$ kN.m

$\sigma_{ULS} = 0.0914 \times 32.10/27.6 = 0.106$

$M_{yT} = 380.6 \times 0.106 = 40.3$ kN.m

Stress due to torsion

$\sigma_{byT} = \dfrac{40.3 \times 10^3}{1668} = 24.2$ N/mm²

Stress due to warping

$\sigma_w = E \, W_{no} \, \theta''$

Commentary to calculation sheet

Commentary to calculation sheet

Some degree of continuity probably exists across the section for Type 1 construction. If this were the case then the effect would be to reduce the applied torsion resulting from the out of balance imposed loads. This degree of continuity would be greatly enhanced if the edge beams could provide some restraint to the floor system as a whole. Edge beams which contain the floor (see Figures 13a, b, c and d) would prevent any horizontal movement of the floor units. This in turn reduces the amount of torsion to the beam.

For obvious reasons the above is difficult to quantify for Type 1 construction. That is why initial assumptions are to ignore any beneficial effects this may have on the system. However, in this case it would seem reasonable to allow for a small percentage in over stress.

The Steel Construction Institute	Job No.	PUB 810		Sheet	18 of 56	Rev.
Silwood Park Ascot Berks SL5 7QN Telephone: (0344) 23345 Fax: (0344) 22944	Job Title	Worked Example No 1				
	Subject	Deflection				
CALCULATION SHEET	Client		Made by DLM		Date Oct 1991	
			Checked by JWR		Date Oct 1991	

Referring to sheet 14 and increasing ϕ'' for the increased loading gives

ϕ'' = $1.50 \times 10^{-8} \times 32.10/27.6 = 1.74 \times 10^{-8}$ Rads

σ_w = $205000 \times 385.8 \times 10^2 \times 1.74 \times 10^{-8}$ = 137.6 N/mm²

Local capacity check

$196 + 24.2 + 137.6 = 357.8$ N/mm² $> p_y = 345$ N/mm² (3.7% over)

<u>Accept for reasons given in commentary.</u>

Case 3b

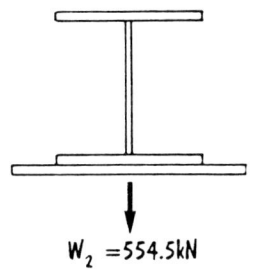

$W_2 = 554.5$ kN

Max moment (Beam restrained)

= $554.5 \times 7.5/8 = 519.8$ kN.m < 817.4 kN.m OK (UF = 0.64)

<u>Case 3 - satisfactory</u>

<u>The combination of torsion and bending for the construction stage and imposed loading conditions are shown to be satisfactory.</u>

DEFLECTIONS

Construction stage - Case 2b

Note, do not use the construction stage loading of 0.5 kN/m² when calculating deflections.

W_2 = $(2.67 + 0.2 + 0.15)7.5^2$

= 170 kN

Deflection due to self weight of pc units and steel beam

$\delta_c = \dfrac{5 \times 170 \times 7500^3}{384 \times 205 \times 41347 \times 10^4} = 11.0$ mm

The Steel Construction Institute	Job No.	PUB 810		Sheet	19	of	56	Rev.
Silwood Park Ascot Berks SL5 7QN Telephone: (0344) 23345 Fax: (0344) 22944	Job Title	Worked Example No 1						
	Subject	Deflection						
CALCULATION SHEET	Client		Made by	DLM		Date	Oct 1991	
			Checked by	JWR		Date	Oct 1991	

Imposed loading - Case 3b

$$W_2 = 3.3 \times 7.5^2 = 185.6 \text{ kN}$$

Deflection due to imposed load

$$\delta_i = \frac{185.6}{170} \times 11.0 = 12.0 \text{ mm } (1/625) \text{ OK}$$

Remaining loads (Superimposed Dead)

Minor services and flooring $0.25 \times 7.5^2 = 14.1$ kN

$$\delta_d = \frac{14.1}{170} \times 11.0 = 1.0 \text{ mm}$$

$$\therefore \text{Total } \delta = 11.0 + 12.0 + 1.0 = 24.0 \text{ mm } (1/312)$$

<u>Vertical deflections considered satisfactory.</u>

The Steel Construction Institute	Job No.	PUB 810		Sheet 20 of 56	Rev.
Silwood Park Ascot Berks SL5 7QN Telephone: (0344) 23345 Fax: (0344) 22944	Job Title	Worked Example No 1			
	Subject	Horizontal Movement of Top Flange for Case 1a			
CALCULATION SHEET	Client		Made by DLM		Date Oct 1991
			Checked by JWR		Date Oct 1991

HORIZONTAL DEFLECTION (TOP FLANGE OF BEAM)

Case 1a - Construction Stage

Using graph 1 on Calculation Sheet 21 $L_t/a = 3.7$

$$\frac{\phi \, G \, J}{T_q \, a} \approx 0.28$$

Unfactored loads

$W_1 = 2.67 \times 7.5^2/2 = 75.1$ kN
$W_2 = 0.2 \times 7.5^2 = 11.3$ kN
(note, do not include the construction loading)
$T_q = 75.1 \times 216/10^3 = 16.2$ kN.m

Rotation of beam

$$\phi = \frac{0.28 \, T_q.a}{G \, J}$$

$$= \frac{0.28 \times 16.2 \times 2031 \times 10^6}{79000 \times 217.3 \times 10^4} = 0.0054 \text{ Rads } (\approx 3°)$$

Horizontal movement of top flange
Distance from shear centre to top flange

$$= D/2 + y + y_0 = \frac{314.5}{2} + 55.7 + 47.9 = 260.9 \text{ mm}$$

∴ horizontal movement $= 260.9 \times 0.054$

$= \underline{14.1 \text{ mm}}$

The max. horizontal movement of the top flange (14.1 mm) occurs at mid span. The main concern is to avoid fouling placement of the pc units and loss of bearing on the units on the bottom plate. Arbitrarily, a 20 mm limit to movement of the top flange would be within normal tolerances.

The Steel Construction Institute	Job No.	PUB 810	Sheet	21	of	56	Rev.
Silwood Park Ascot Berks SL5 7QN Telephone: (0344) 23345 Fax: (0344) 22944	Job Title	Worked Example No 1					
	Subject	Graphs 1 & 2 used for Rigorous Analysis					
CALCULATION SHEET	Client		Made by DLM			Date Oct 1991	
			Checked by JWR			Date Oct 1991	

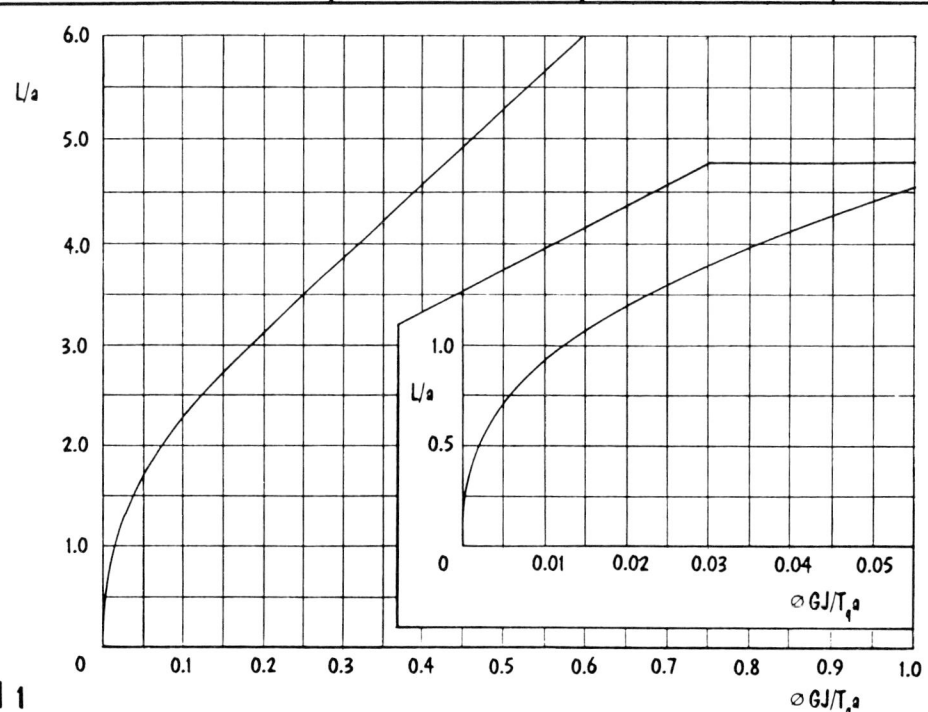

GRAPH 1

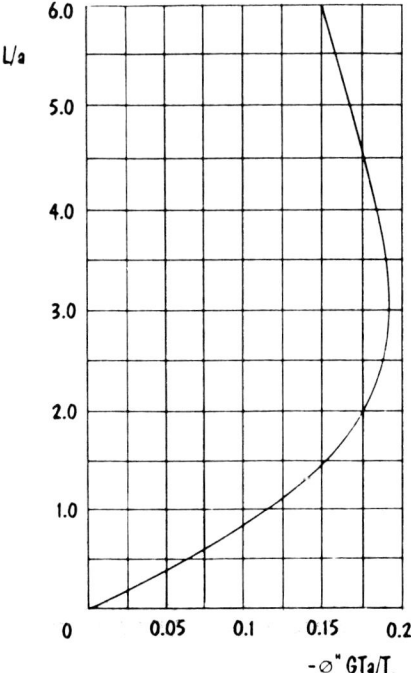

GRAPH 2

Note: These two graphs have been modified from the graphs shown in Reference 6.

The Steel Construction Institute	Job No. **PUB 810**		Sheet **22** of **56**	Rev.
Silwood Park Ascot Berks SL5 7QN Telephone: (0344) 23345 Fax: (0344) 22944	Job Title **Worked Example No 1**			
	Subject **Serviceability Stresses**			
CALCULATION SHEET	Client	Made by **DLM**	Date **Oct 1991**	
		Checked by **JWR**	Date **Oct 1991**	

SERVICEABILITY STRESSES
Case 3b

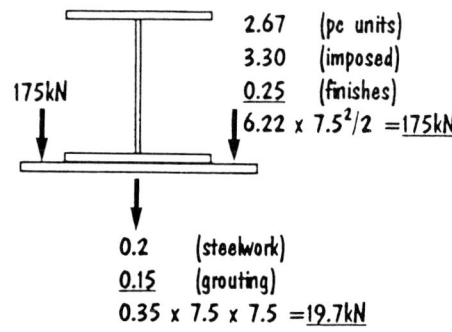

2.67 (pc units)
3.30 (imposed)
0.25 (finishes)
6.22 × 7.5²/2 = 175kN

175kN

0.2 (steelwork)
0.15 (grouting)
0.35 × 7.5 × 7.5 = 19.7kN

Total load at servicability limit state
= 175 + 175 + 19.7
= 369.7 kN

Max moment = 369.7 × 7.5/8 = 346.6 kN.m

I_{xx} = 41347 cm^4

Z_c = $\frac{41347}{21.3}$ = 1941 cm^3 Z_t = $\frac{41347}{11.66}$ = 3546 cm^3

Compressive stress in top flange

= $\frac{346.6 \times 10^3}{1941}$ = 178.6 N/mm^2 < p_y = 345 N/mm^2 OK

The combination of the extreme tensile fibre stress with the transverse bending stress is not a critical design case for this scheme (for further information see worked example No. 3).

	Job No.	PUB 810	Sheet 23 of 56	Rev.
The Steel Construction Institute	Job Title	Worked Example No 1		
Silwood Park Ascot Berks SL5 7QN Telephone: (0344) 23345 Fax: (0344) 22944	Subject	Vertical Shear		
CALCULATION SHEET	Client	Made by DLM	Date Oct 1991	
		Checked by JWR	Date Oct 1991	

VERTICAL SHEAR

Case 3b

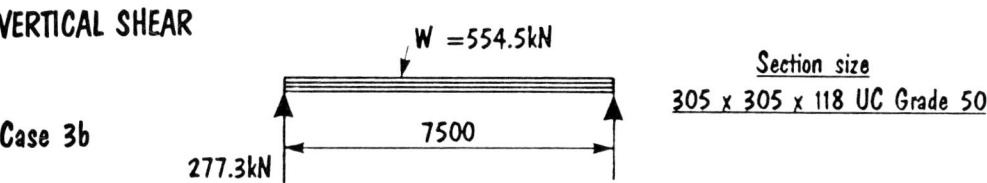

Section size
305 x 305 x 118 UC Grade 50

∴ Shear force F_v = 277.3 kN

Shear capacity, P_v

$P_v = 0.6\, p_y \cdot A_v$ where: $A_v = t_w D$
                                         $= 11.9 \times 314.5$

$= \dfrac{0.6 \times 345 \times 11.9 \times 314.5}{10^3}$

$= 774.7$ kN $>$ 277.3 kN   OK

<u>Vertical shear satisfactory</u>

Note: No need to check for combination of moment and high shear $(0.6P_v)$ as applied loads are uniformly distributed.

	Job No.	PUB 810	Sheet 24 of 56	Rev.
The Steel Construction Institute	Job Title	Worked Example No 1		
Silwood Park Ascot Berks SL5 7QN Telephone: (0344) 23345 Fax: (0344) 22944	Subject	Natural Frequency		
CALCULATION SHEET	Client	Made by DLM	Date Oct 1991	
		Checked by JWR	Date Oct 1991	

NATURAL FREQUENCY

Loading = Dead + 10% Imposed (not including partitions)

Dead = (2.67 + 0.25 + 0.2 + 0.15) = 3.27 kN/m²
Imposed = 10% of 2.4 kN/m² = 0.24 kN/m²
                                  3.51 kN/m²

∴ Total Load = 3.51 × 7.5 × 7.5
             = 197.4 kN

Deflection due to permanent loads

$$\delta_{sw} = \frac{5 \times 197.4 \times 7500^3}{384 \times 205 \times 41347 \times 10^4}$$

= 12.8 mm

Natural frequency $\approx \dfrac{18}{\sqrt{\delta_{sw}}} = \dfrac{18}{\sqrt{12.8}}$

= 5.0 H_z > 4.0 H_z limit OK

∴ **Floor structure satisfactory for natural frequency**

The Steel Construction Institute	Job No. PUB 810	Sheet 25 of 56	Rev.
Silwood Park Ascot Berks SL5 7QN Telephone: (0344) 23345 Fax: (0344) 22944	Job Title Worked Example No 2		
	Subject Loading		
CALCULATION SHEET	Client	Made by DLM	Date Oct 1991
		Checked by JWR	Date Oct 1991

TYPE 2 CONSTRUCTION

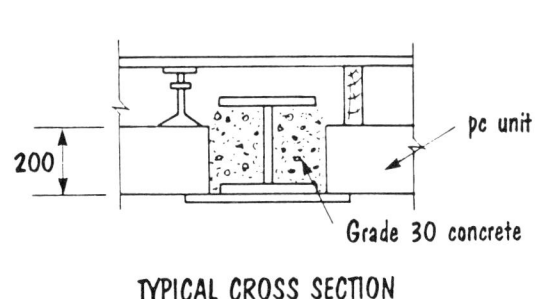

TYPICAL CROSS SECTION
THROUGH BEAM

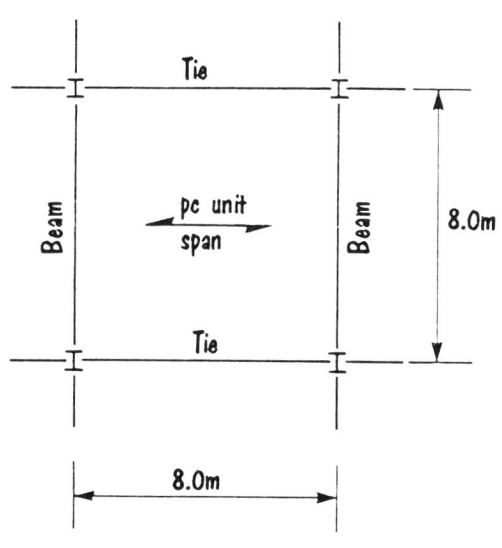

LOADING - DEAD kN/m^2
PC units   2.67
Self wt of steel   0.20
Construction load   0.50
Flooring & services, Say   0.40
Beam encasement   0.30
Ceiling,   Say   0.20

LOADING - IMPOSED
Occupancy   3.5
Partitions   <u>1.0</u>
           <u>4.5</u>

Imposed load reduction
supported area = $8 \times 8 = 64\ m^2$

Reduction
= $4.5 \times 0.064 = 0.3\ kN/m^2$

∴ Revised Imposed Load

= 4.5 - 0.3

= $4.2\ kN/m^2$

The Steel Construction Institute	Job No. PUB 810	Sheet 26 of 56	Rev.
Silwood Park Ascot Berks SL5 7QN Telephone: (0344) 23345 Fax: (0344) 22944	Job Title Worked Example No 2		
	Subject Load Combinations		
CALCULATION SHEET	Client	Made by DLM	Date Oct 1991
		Checked by JWR	Date Oct 1991

LOADING COMBINATIONS (from Worked Example No. 1)

To determine the design criterion for bending:

Case 1a — The difference between schemes 1 and 2 is that the span has gone up to 8 m from 7.5 m. For the purposes of this example consider beam is OK for Case 1.
Case 2 — Consider beam is OK for Case 2.
Case 3 — Beam considered to be laterally restrained, check for max. factored moment

∴ Design for Case 3b

Note: Cases 1 and 2 have not been shown because the procedures for these design cases are covered in the first worked example. In addition, Case 3a does not apply to Types 2 and 3 forms of construction. See note (4) in general assumptions.

Firstly, check biaxial stress effects in the plate using graph shown in Figure 23

Case 3b

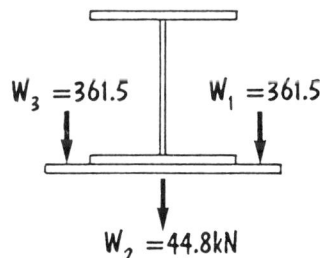

$W_3 = 361.5$ $W_1 = 361.5$

$W_2 = 44.8$ kN

$W_1 = [(2.67 + 0.4 + 0.2) 1.4 + (4.2 \times 1.6)]8^2/2$ = 361.5 kN
$W_2 = (0.2 + 0.3) 1.4 \times 8^2$ = 44.8 kN

Plate moment = $361.5 \times 62.5/10^3$ = 22.6 kN.m
Moment per unit length = 22.6/8 = 2.82 kN.m/m

$S_{xp} = 1.5^2 \times 800/4 = 450$ cm³
Section modulus per unit length = 450/8 = 56.25 cm³/m
Moment capacity = $56.25 \times 355/10^3$ = 19.97 kN.m/m

∴ Actual plate moment ratio = 2.82/19.97 = 0.14

Total vertical factored loads = 361.5 + 361.5 + 44.8 = 767.8 kN
Max. moment applied to beam = $767.8 \times 8/8$ = 767.8 kN.m

		Job No.	PUB 810	Sheet 27 of 56	Rev.
The Steel Construction Institute 		Job Title	Worked Example No 2		
Silwood Park Ascot Berks SL5 7QN Telephone: (0344) 23345 Fax: (0344) 22944		Subject	Deflection		
		Client		Made by DLM	Date Oct 1991
CALCULATION SHEET				Checked by JWR	Date Oct 1991

LONGITUDINAL BENDING STRESS, σ_1

$$\sigma_1 = \frac{767.8 \times 10^6}{2369.3 \times 10^3} = 324.1 \text{ N/mm}^2$$

$$\frac{\sigma_1}{p_y} = \frac{324.1}{355} = 0.913$$

from the graph shown in Figure 23 using $\frac{\sigma_1}{p_y} = 0.913$, $\frac{M}{M_p} \approx 0.27$

<u>As 0.27 > 0.14   OK</u>

∴ No influence on overall bending due to biaxial stresses

Check maximum factored moment

Applied moment   =   767.8 kN.m (see above)
Moment capacity   =   817.4 kN.m > 767.8 kN.m   OK

∴ <u>Construction and Imposed Load Bending Capacity satisfactory.</u>

DEFLECTION

Imposed load

$$W = 8 \times 8 \times 4.2 = 268.8 \text{ kN}$$

Area of concrete divided by modular ratio of 10

$$A_c = \frac{30.68 \,(31.45 - 2 \times 1.87)}{10} = 85.0 \text{ cm}^2$$

Modified neutral axis position from mid-height of beam

$$\bar{y} = \frac{76.5 \,(314.5 + 15)}{2(226.3 + 85.0)} = 40.5 \text{ mm}$$

	Job No.	PUB 810	Sheet 28 of 56	Rev.
The Steel Construction Institute Silwood Park Ascot Berks SL5 7QN Telephone: (0344) 23345 Fax: (0344) 22944 **CALCULATION SHEET**	Job Title	Worked Example No 2		
	Subject	Uncracked Inertia		
	Client	Made by DLM	Date Oct 1991	
		Checked by JWR	Date Oct 1991	

UNCRACKED INERTIA

Consider combined second moment of area of steel section and concrete encasement

$$I_{xx} = (27601 + 149.8 \times 4.05^2) + 76.5(16.47 - 4.05)^2 + \frac{30.68(31.45 - 2 \times 1.87)^3}{12 \times 10} + 85.0 \times 4.05^2$$

$$= 48693 \text{ cm}^4$$

Deflection of beam subject to imposed load

$$\delta_i = \frac{5 \times 268.8 \times 8000^3}{384 \times 205 \times 48693 \times 10^4} = 18.0 \text{ mm } (1/444) \text{ OK}$$

	Job No.	PUB 810		Sheet 29 of 56	Rev.
The Steel Construction Institute	Job Title	Worked Example No 2			
Silwood Park Ascot Berks SL5 7QN Telephone: (0344) 23345 Fax: (0344) 22944	Subject	Serviceability Stresses			
	Client		Made by DLM		Date Oct 1991
CALCULATION SHEET			Checked by JWR		Date Oct 1991

SERVICEABILITY STRESSES (Steel beam only)

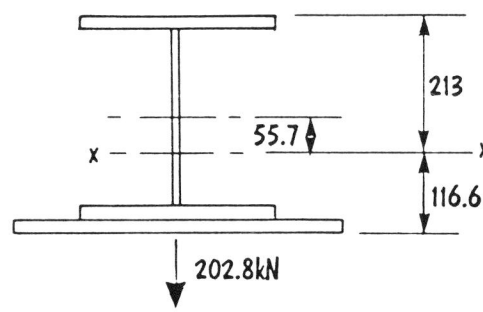

Loading

pc units = 2.67 × 8 × 8 = 170.8 kN

steel & concrete = (0.2 + 0.3) 8 × 8 = 32.0 kN
                                       202.8 kN

Moment due to dead loads = 202.8 × 8/8 = 202.8 kN.m

Longitudinal stresses

$$\sigma_c = \frac{202.8 \times 10^3}{1941} = 104.5 \text{ N/mm}^2$$

$$\sigma_t = \frac{202.8 \times 10^3}{3546} = 57.2 \text{ N/mm}^2$$

SERVICEABILITY STRESSES (Encased beam)

The stresses in the steel and concrete may be determined using the following approach. It is first necessary to calculate the elastic properties of the encased section. This is the rigorous approach used in the computer software. Designers may also ignore the increased stiffness of the encased beam, in which case the serviceability stress checks are as for Type 1 construction.

Unfactored loads applied to encased section

Imposed = 4.2 kN/m²
Flooring & Services = 0.4 kN/m²
Ceiling = 0.2 kN/m²
      4.8 kN/m²

The Steel Construction Institute	Job No. **PUB 810**	Sheet **30** of **56**	Rev.
Silwood Park Ascot Berks SL5 7QN Telephone: (0344) 23345 Fax: (0344) 22944	Job Title **Worked Example No 2**		
	Subject **Serviceability Stresses cont'd**		
CALCULATION SHEET	Client	Made by **DLM**	Date **Oct 1991**
		Checked by **JWR**	Date **Oct 1991**

Total load = 4.8 × 8 × 8 = 307.2 kN,
Moment = 307.2 × 8/8 = 307.2 kN.m

Cracked section properties

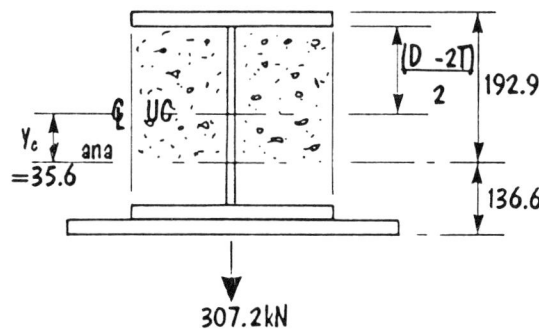

Cracked section properties used for serviceability stress calculations are from Appendix B, Case (2b) the downward dimension to the n.a (y) is given by:

$$y_c^2 (k_1) + y_c (k_2) + (k_3) = 0$$

where $k_1 = \dfrac{B}{2\alpha_e} = \dfrac{306.8}{2 \times 10} = 15.34$

α_e = modular ratio = 10

$k_2 = A_T + \dfrac{B}{\alpha_e}\left(\dfrac{D - 2T}{2}\right)$

$= 226.3 \times 10^2 + \dfrac{306.8}{10}\left(\dfrac{314.5 - 2 \times 18.7}{2}\right) = 26881$

$k_3 = \dfrac{B}{2\alpha_e}\left(\dfrac{D - 2T}{2}\right)^2 - A_p\left(\dfrac{D + t_p}{2}\right)$

$= \dfrac{306.8}{2 \times 10}\left(\dfrac{314.5 - 2 \times 18.7}{2}\right)^2 - 7650\left(\dfrac{314.5 + 15}{2}\right)$

$= -965869$

	Job No.	PUB 810	Sheet 31 of 56	Rev.
The Steel Construction Institute	Job Title	Worked Example No 2		
Silwood Park Ascot Berks SL5 7QN Telephone: (0344) 23345 Fax: (0344) 22944	Subject	Cracked Inertia		
CALCULATION SHEET	Client	Made by DLM	Date Oct 1991	
		Checked by JWR	Date Oct 1991	

hence $y_c = \dfrac{-26881 \pm \sqrt{26881^2 + 4 \times 15.34 \times 965869}}{2 \times 15.34}$

$\therefore y_c = 35.6$ mm (elastic neutral axis below mid height of UC)

Concrete area (in steel units) from Appendix B

$A_c = \dfrac{B}{\alpha_e} [(D-2T)/2 + y_c]$

$= \dfrac{306.8}{10} \left[\left(\dfrac{314.5 - 2 \times 18.7}{2} \right) + 35.6 \right]$

$= 5343$ mm^2

Second moment of area, I_x from Appendix B is given by:

$I_x = k_4 + k_5 + k_6 + k_7$

where:

$k_4 = I_{uc} + Ay_c^2 = 27601 + 149.8 \times 3.56^2 = 29500$ cm^4

$k_5 = A_p \left(\dfrac{D + t_p}{2} - y_c \right)^2 = 76.5 \left(\dfrac{314.5 + 15}{20} - 3.56 \right)^2 = 12760$ cm^4

$k_6 = \dfrac{B}{12\alpha_e} \left[\left(\dfrac{D-2T}{2} \right) + y_c \right]^3 = \dfrac{30.68}{12 \times 10} \left[\left(\dfrac{314.5 - 2 \times 18.7}{20} \right) + 3.56 \right]^3$

$= 1350$ cm^4

The Steel Construction Institute	Job No.	PUB 810		Sheet 32 of 56	Rev.
Silwood Park Ascot Berks SL5 7QN Telephone: (0344) 23345 Fax: (0344) 22944	Job Title	Worked Example No 2			
	Subject	Combined Steel Serviceability Stresses			
CALCULATION SHEET	Client		Made by DLM	Date	Oct 1991
			Checked by JWR	Date	Oct 1991

$$k_7 = A_c \left[\frac{\left(\frac{D-2T}{2}\right) - y_c}{2}\right]^2 = 53.43 \left[\frac{\left(\frac{314.5 - 2 \times 18.7}{20}\right) - 3.56}{2}\right]^2 = 1416 \text{ cm}^4$$

$$\therefore I_x = 29500 + 12760 + 1350 + 1416 = 45026 \text{ cm}^4$$

$$Z \text{ (concrete)} = \frac{I_x \cdot \alpha_e}{\left(\frac{D-2T}{2} + y_c\right)} = \frac{45026 \times 10}{\left(\frac{314.5 - 2 \times 18.7}{20} + 3.56\right)} = 25855 \text{ cm}^3$$

Max concrete stress $= \dfrac{307.2 \times 10^2}{25855} = 11.9 \text{ N/mm}^2 < \dfrac{30}{2} = 15 \text{ N/mm}^2$ $\quad \therefore$ OK

Steel stesses

$$Z_c = \frac{45026}{19.29} = 2334 \text{ cm}^3; \quad Z_t = \frac{45026}{13.66} = 3296 \text{ cm}^3$$

$$\sigma_c = \frac{307.2 \times 10^3}{2334} = 131.6 \text{ N/mm}^2$$

$$\sigma_t = \frac{307.2 \times 10^3}{3296} = 93.2 \text{ N/mm}^2$$

Combined steel stresses for steel beam and encased section

$$\sigma_c = 104.5 + 131.6 = 236.1 \text{ N/mm}^2 < 345 \text{ N/mm}^2$$

$$\sigma_t = 57.2 + 93.2 = 150.4 \text{ N/mm}^2$$

$\therefore$ Serviceability stresses for steel and concrete satisfactory.

	Job No.	PUB 810	Sheet 33 of 56	Rev.
The Steel Construction Institute	Job Title	Worked Example No 2		
Silwood Park Ascot Berks SL5 7QN Telephone: (0344) 23345 Fax: (0344) 22944	Subject	Natural Frequency		
	Client		Made by DLM	Date Oct 1991
CALCULATION SHEET			Checked by JWR	Date Oct 1991

The previous calculations show that the increased second moment of area of the encased section relative to the steel section is about 20%.

NATURAL FREQUENCY

Dead   = 2.67 + 0.2 + 0.4 + 0.3 + 0.2 = 3.77 kN/m²
Imposed = 0.1 × 3.28   = 0.33 kN/m²
                                                             4.10 kN/m²

Permanent Load = 4.10 × 8 × 8
                       = 262.4 kN

Deflection of encased beam

$$\delta_{sw} = \frac{5 \times 262.4 \times 8000^3}{384 \times 205 \times 48693 \times 10^4}$$

= 17.5 mm

Natural frequency

$$f = \frac{18}{\sqrt{17.5}}$$

= 4.3 H_z > 4.0 H_z OK

TYPE 3 CONSTRUCTION

LOADING - CONSTRUCTION STAGE

	kN/m²
PC units	4.0
Conc. encasement	0.3
Self wt. of steel	0.2
	4.5
Construction Load	0.5

Dead composite
Self wt. of in situ slab	3.2
Ceiling & Services	0.4
	3.6

LOADING - IMPOSED
Occupancy	5.0
Partitions	1.0
	6.0

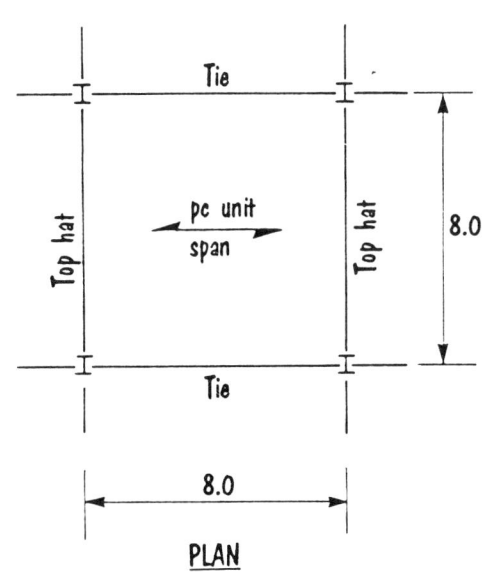

PLAN

Note: Where the dead loads are considered high, an alternative option may be to use lightweight concrete.

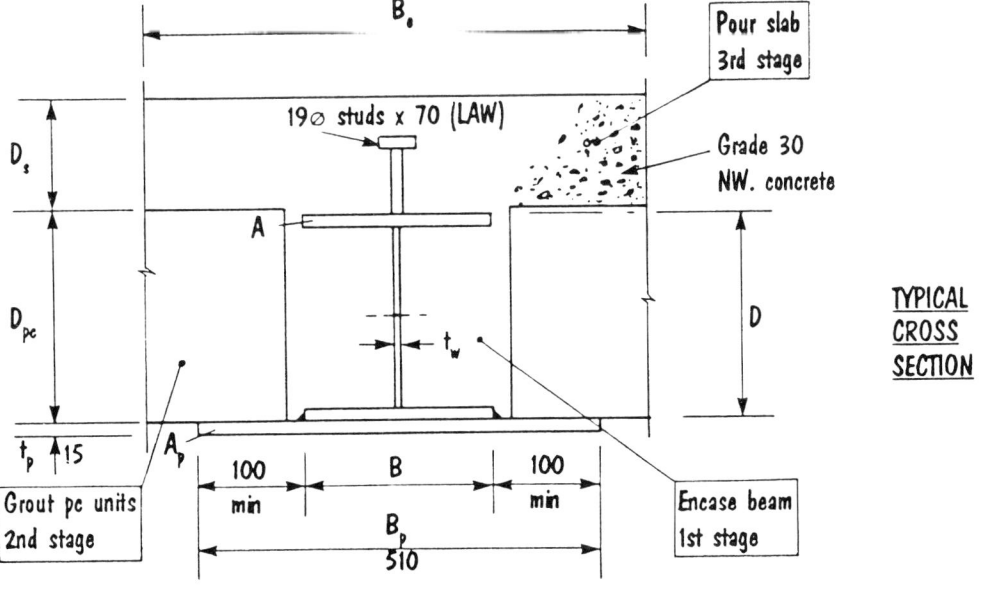

TYPICAL CROSS SECTION

Supported area = 64 m²
Imposed reduction = 6 × 0.064 = 0.4 kN/m²
∴ Imposed load = 6 − 0.4 = 5.6 kN/m²

Note: The procedures for the initial stages of construction are covered in the first worked example.

The Steel Construction Institute Silwood Park Ascot Berks SL5 7QN Telephone: (0344) 23345 Fax: (0344) 22944 **CALCULATION SHEET**	Job No. **PUB 810**	Sheet **35** of **56** Rev.
	Job Title **Worked Example No 3**	
	Subject **Factored Loading**	
	Client	Made by **DLM** Date **Oct 1991**
		Checked by **JWR** Date **Oct 1991**

Steel Beam : Try 305 × 305 × 118 UC Grd. 50
Plate   : 510 × 15 thick Grd. 50

The beam is laterally restrained and is designed to act compositely to resist the total factored load.

Note: If stages 1 and 3 are done in one operation then the beam would have to be considered as laterally unrestrained.

COMPOSITE

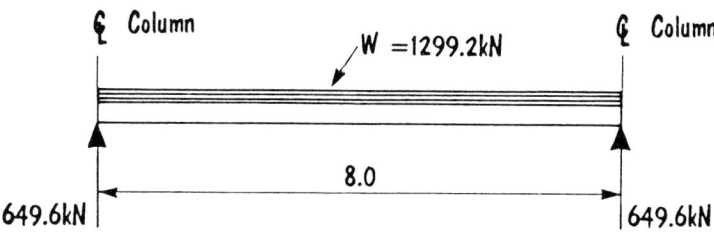

TOTAL FACTORED LOADS [Case 3b]

$W = (4.5 + 3.6) \, 1.4 + 5.6 \times 1.6 \quad = 20.3 \text{ kN/m}^2$

$W = 20.3 \times 8 \times 8 \quad = 1299.2 \text{ kN}$

Max. moment $= 1299.2 \times 8/8$

$= 1299.2 \text{ kN.m}$

The Steel Construction Institute	Job No. **PUB 810**		Sheet **36** of **56**	Rev.
Silwood Park Ascot Berks SL5 7QN Telephone: (0344) 23345 Fax: (0344) 22944	Job Title **Worked Example No 3**			
	Subject **Case 1a Construction stage - Torsion**			
CALCULATION SHEET	Client	Made by **DLM**	Date **Oct 1991**	
		Checked by **JWR**	Date **Oct 1991**	

CONSTRUCTION STAGE CASE 1a

pc units 4.0 kN/mm²
steelwork 0.2 kN/mm²
construction 0.5 kN/mm²

8m × 8m

Check to see if a full internal bay can be erected with the pc units placed on one side only.

```
4 × 1.4 × 8 × 8/2   =  179.2
0.5 × 1.6 × 8 × 8/2 =   25.6
                      204.8 kN
```

↓ 0.2 × 8 × 8 × 1.4 = 17.9 kN

$M_1 = 204.8 \times \dfrac{8}{8} = 204.8$

$M_2 = 17.9 \times \dfrac{8}{8} = \underline{17.9}$

$M_x = \underline{22.7 \text{ kN.m}}$

Equivalent transverse force in flanges to resist torsion

$F = 204.8 \times \dfrac{216}{305} = 145 \text{ kN}$

$M_3 = 145 \times \dfrac{8}{8} = 145 \text{ kN.m}$

CONSTRUCTION STAGE – Case 1a continued

Transverse moment in top flange

$M_{cy} = 151.8$ kN.m

Unity factor (UF) $= \dfrac{222.7}{635} + \dfrac{145}{151.8} = 1.31$ ∴ erection of the pc units will have to be controlled in the construction stage.

Try z = 5 m

Note:

The dimension z is the extent the pc units can be erected before adjacent bays of floor units have to be laid to restore equilibrium to the beam.

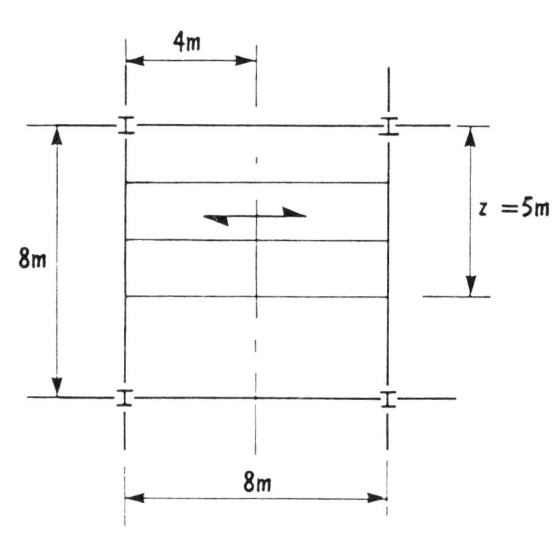

LOADING

$4 \times 1.4 \times 4 \times 5 = 112.0$ kN
$0.5 \times 1.6 \times 4 \times 5 = \underline{16.0 \text{ kN}}$
$ \underline{128.0 \text{ kN}}$

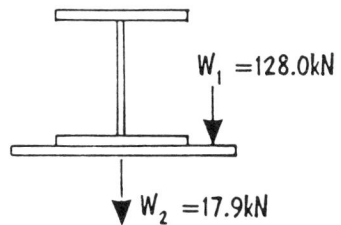

$M_1 = \dfrac{128.0 \times 5}{2}\left(1 - \dfrac{5}{2 \times 8}\right)^2 = 151.3$ kN.m

$M_2 = \dfrac{17.9 \times 3.44}{2 \times 8}(8 - 3.44) = 17.6$ kN.m

$x_1 = 5\left(1 - \dfrac{5}{2 \times 8}\right) = 3.44$ m

$M_x = 151.3 + 17.6 = 168.9$ kN.m

The Steel Construction Institute		Job No.	PUB 810	Sheet **38** of **56**	Rev.
Silwood Park Ascot Berks SL5 7QN Telephone: (0344) 23345 Fax: (0344) 22944		Job Title	Worked Example No 3		
		Subject	Biaxial Stress Effects		
CALCULATION SHEET		Client	Made by DLM	Date Oct 1991	
			Checked by JWR	Date Oct 1991	

$$F = 128 \times \frac{216}{305} = 90.6 \text{ kN}$$

$$M_y = \frac{90.6 \times 5}{2}\left(1 - \frac{5}{2 \times 8}\right)^2 = 107.0 \text{ kN.m}$$

$$\frac{168.9}{635} + \frac{107}{151.8} = 0.97 < 1.0 \quad \therefore z = 5 \text{ m}$$

Check combined logitudinal stresses with transverse stress to see if a reduction in p_y is required. Conservatively, all loads apart from the self weight of steel and encased concrete will be assumed to act as an eccentric load.

Calculate transverse bending ratio, $\frac{M}{M_p}$

Loading

$$(4 + 3.2 + 0.4)\,1.4 \times 8 \times \frac{8}{2} = 340.5 \text{ kN}$$

$$5.6 \times 1.6 \times 8 \times \frac{8}{2} = \underline{286.7 \text{ kN}}$$

$$\underline{627.2 \text{ kN}}$$

Plate moment, $M = 627.2 \times 62.5/10^3$

$$= 39.2 \text{ kN.m}$$

$$M_p = \frac{8000 \times 15^2 \times 355}{4 \times 10^6} = 159.8 \text{ kN.m}$$

$$\therefore \frac{M}{M_p} = \frac{39.2}{159.8} = 0.245, \quad \text{from Figure 23 } \frac{\sigma_1}{p_y} \approx 0.93$$

$$\therefore \sigma_1 = 0.93 \times 355 = 330 \text{ N/mm}^2$$

This is now the design strength of the bottom plate.

	Job No.	PUB 810	Sheet 39 of 56	Rev.
The Steel Construction Institute	Job Title	Worked Example No 3		
Silwood Park Ascot Berks SL5 7QN Telephone: (0344) 23345 Fax: (0344) 22944	Subject	Composite Section		
CALCULATION SHEET	Client		Made by DLM	Date Oct 1991
			Checked by JWR	Date Oct 1991

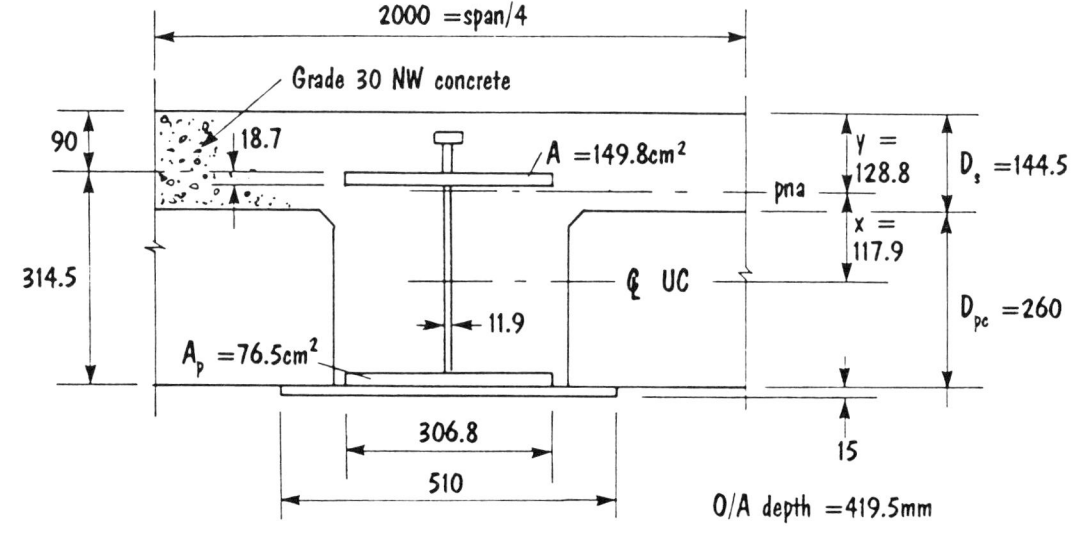

Steel Beam 305 × 305 × 118 UC   Grd. 50   Area = 149.8 cm²
Plate   510 × 15 thick   Grd. 50   Area = 76.5 cm²
                                                                                                                                                                       226.3 cm²

Step 1   <u>Calc. y for full cross-section, distance to the pna from top of slab</u>

R_s = 149.8 × 345/10 = 5168.1 kN
R_p = 76.5 × 330/10 = 2524.5 kN
                                                                                                             7692.6 kN   ($R_s + R_p$)

Try CASE 2 FS   See Appendix B

$$x = \frac{0.45 \times 30 \times 2000/10^3 \,(144.5 + 260 - 314.5/2) - 2524.5}{0.45 \times 30 \times 2000/10^3 + 2 \times 345 \times 11.9/10^3}$$

  = 117.9 mm

y = 144.5 + 260 - 314.5/2 - 117.9 = <u>128.8 mm</u>

∴ position y, assumed in Web - CASE 2 FS   OK

Step 2   <u>Calc. R_c' for full cross-section</u>

R_c' = 0.45 × 30 × 2000 × 128.8/10³ = <u>3493.8 kN</u>

The Steel Construction Institute	Job No.	PUB 810	Sheet **40** of **56**	Rev.
Silwood Park Ascot Berks SL5 7QN Telephone: (0344) 23345 Fax: (0344) 22944	Job Title	Worked Example No 3		
	Subject	Shear Stud Connector Design		
CALCULATION SHEET	Client	Made by DLM	Date Oct 1991	
		Checked by JWR	Date Oct 1991	

Step 3 Compare R_c' with $(R_s + R_p)$ and use the lesser of these two values

R_c' = 3493.8 < 7807.4 kN

As R_c' < $(R_s + R_p)$ determine number of shear connectors

using R_c' = <u>3493.8 kN</u>

Step 4 Shear Stud Connector Design

Stud capacity $19\phi \times 70$ (LAW)

Q_k = 87 kN (Grd. 30 NW concrete)

Q_p = 0.8 Q_k = 0.8 × 87 = 69.6 kN

N_p = F_p/Q_p = R_c'/Q_p = 3493.8/69.6 = 50.2 Say 51

Total number of studs for full shear connection
= 51 × 2 = 102 studs

Step 5 Partial Shear Connection

Assume R_q = 0.4 R_c' = 0.4 × 3493.8 = 1397.5 kN

Step 6 Moment Capacity, M_c

Note: 40% is the minimum amount of shear connection permitted in BS 5950: Part 3: Section 3.1.

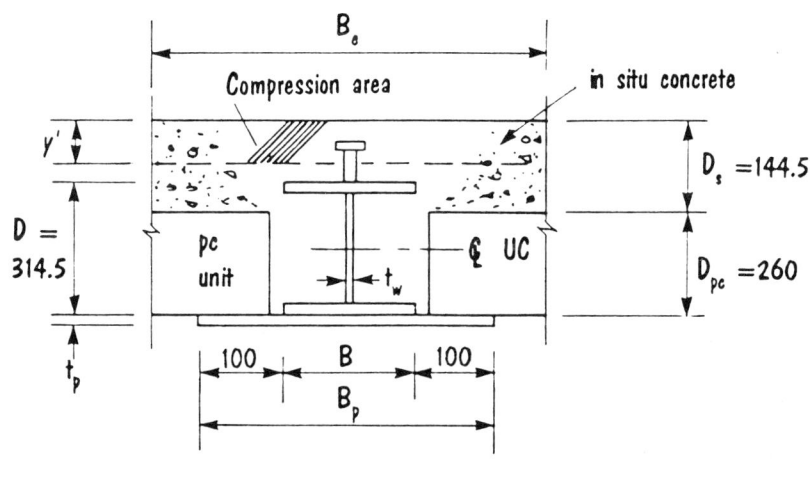

	Job No.	PUB 810	Sheet 41 of 56	Rev.
The Steel Construction Institute  Silwood Park Ascot Berks SL5 7QN Telephone: (0344) 23345 Fax: (0344) 22944	Job Title	Worked Example No 3		
	Subject	Moment Capacity		
CALCULATION SHEET	Client	Made by DLM	Date Oct 1991	
		Checked by JWR	Date Oct 1991	

a) by trial and error using the four equations find x.

try pna in the web below centreline of UC, see Case 3PS in Appendix B

R_p = 2524.5 kN

R_q = 1397.5 kN $x = \dfrac{R_p - R_q}{2 p_y t_w} = \dfrac{2524.5 - 1397.5}{2 \times 345 \times 11.9/10^3}$

$= 137.3$ mm

∴ Assumption correct, pna in web of UC below centre line off UC

$$M_c = M_s + R_q \left(D_{pc} + D_s - \dfrac{D}{2} - \dfrac{R_q}{0.9 f_{cu} B_e} \right) + \dfrac{R_p}{2} (D + t_p) - \dfrac{(R_p - R_q)^2}{4 p_y t_w}$$

b) check $y' < y'_{crt}$ to ensure sufficient concrete available

$y' = R_q/0.45 f_{cu}.B_e = \dfrac{1397.5 \times 10^3}{0.45 \times 30 \times 2000} = 51.76$ mm

$D - 50 = 314.5 - 50 = 264.5$ mm, $D_{pc} = 260$ mm < 264.5 mm

from Appendix B (cases for partial shear connection) since $D_{pc} < (D - 50)$.

$y_{crt} = D_s + D_{pc} - D + 50$

$= 144.5 + 260 - 314.5 + 50$

$= 140$ mm

Since 140 mm $>$ 51.76 mm y' lies within limit.

$$M_c = \dfrac{\dfrac{1950 \times 10^3 \times 345}{10^6} + 1397.5}{10^3} \left(260 + 144.5 - \dfrac{314.5}{2} - \dfrac{1397.5 \times 10^3}{0.9 \times 30 \times 2000} \right) + \dfrac{2524.5}{2 \times 10^3} (314.5 + 15) - \dfrac{(2524.5 - 1397.5)^2}{4 \times 345 \times 11.9}$$

$= 672.8 + 309.4 + 415.9 - 77.3$

$= 1320.8$ kN.m > 1299.2 kN.m OK

The Steel Construction Institute		Job No.	PUB 810		Sheet 42 of 56	Rev.
Silwood Park Ascot Berks SL5 7QN Telephone: (0344) 23345 Fax: (0344) 22944		Job Title	Worked Example No 3			
		Subject	Number of Studs and Layout			
CALCULATION SHEET		Client		Made by DLM	Date Oct 1991	
				Checked by JWR	Date Oct 1991	

Step 7 — **Determine Number of Studs for Partial Shear Connection**

Using R_q = 1397.5 kN the above value for $M_c > M_{applied}$

∴ number of studs = 0.4 × 50.2 = 20.1 <u>say 21</u>

or $\dfrac{1397.5}{69.6}$ = 20.1

Stud Layout

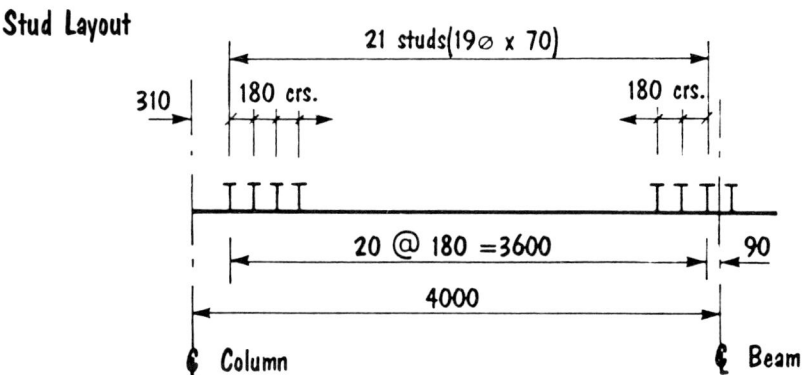

Minimum spacing = S_d = 5 × 19 = 95 mm < 180 mm OK

The Steel Construction Institute	Job No.	PUB 810	Sheet 43 of 56	Rev.
Silwood Park Ascot Berks SL5 7QN Telephone: (0344) 23345 Fax: (0344) 22944	Job Title	Worked Example No 3		
	Subject	Deflection		
CALCULATION SHEET	Client		Made by DLM	Date Oct 1991
			Checked by JWR	Date Oct 1991

Step 8   **Deflection**

As partial shear connection has been used the deflections will have to be increased due to the slippage of the stud connectors.

For unpropped construction $= \delta_c + 0.3\left(1 - \dfrac{N_a}{N_p}\right)(\delta_s - \delta_c)$

δ_s = steel beam acting alone.
δ_c = composite beam with full shear connection.

$\dfrac{N_a}{N_p}$ taken as 0.4

a)   Section Properties

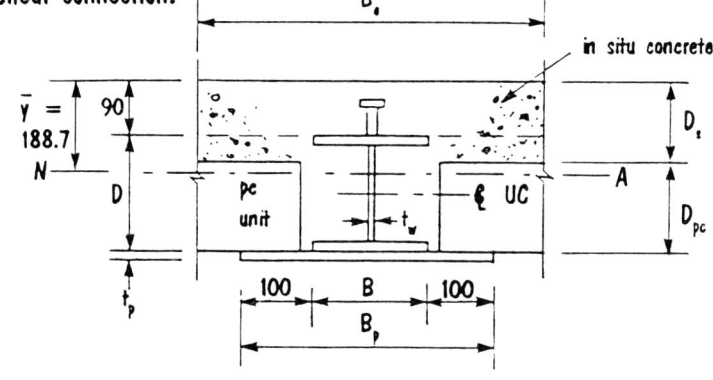

Steel section $I_{xx} = 41347 \text{ cm}^4$

Composite Section

$$\bar{y} = \dfrac{\dfrac{2000}{2 \times 10} \times 90^2 + 149.8 \times 10^2(404.5 - 157.25) + 76.5 \times 10^2(404.5 + 15/2)}{(76.5 + 149.8)10^2 + \dfrac{2000}{10} \times 90}$$

$= 188.7$ mm

Second Moment of Area, I_{NA}

$K_1 = 188.7 - 90/2 = 143.7$ mm $K_2 = 404.5 - \dfrac{314.5}{2} - 188.7 = 58.6$ mm

$K_3 = 404.5 + 7.5 - 188.7 = 223.3$ mm

	Job No.	PUB 810	Sheet 44 of 56	Rev.
The Steel Construction Institute Silwood Park Ascot Berks SL5 7QN Telephone: (0344) 23345 Fax: (0344) 22944	Job Title	Worked Example No 3		
	Subject	Deflection cont'd		
CALCULATION SHEET	Client	Made by DLM	Date Oct 1991	
		Checked by JWR	Date Oct 1991	

$$I_x = \frac{2000 \times 90}{10 \times 100}\left(\frac{9^2}{12} + 14.37^2\right) + 27601 + 149.8 \times 5.86^2 + 76.5 \times 22.33^2$$

$$= 38384 + 27601 + 5144 + 38145$$

$$= 109274 \text{ cm}^4$$

b) **Imposed Load Deflection**

$$W = 8 \times 8 \times 5.6 = 358.4 \text{ kN}$$

$$\delta_s = \frac{5}{384} \times \frac{358.4 \times 8000^3}{205 \times 41347 \times 10^4} = 28.2 \text{ mm}$$

$$\delta_c = 28.2 \times 41347/109274 = 10.7 \text{ mm}$$

$$\delta = 10.7 + 0.3 (1 - 0.40)(28.2 - 10.7)$$

$$= 10.7 + 3.2 \text{ mm}$$

$$= 13.9 \text{ mm } (1/576) < (L/360) \text{ OK}$$

∴ Imposed deflection satisfactory

c) **Dead Load Deflection (Construction stage)**

Loading $= 7.7 \times 8 \times 8 = 492.8 \text{ kN}$

$$\delta_d = \frac{492.8}{358.4} \times 28.2 = 38.8 \text{ mm}$$

The Steel Construction Institute		Job No.	**PUB 810**		Sheet **45** of **56**	Rev.
Silwood Park Ascot Berks SL5 7QN Telephone: (0344) 23345 Fax: (0344) 22944		Job Title	Worked Example No 3			
		Subject	Serviceability Stresses			
CALCULATION SHEET		Client		Made by DLM		Date Oct 1991
				Checked by JWR		Date Oct 1991

SERVICEABILITY STRESSES

All loads except for the self weight of steel and encased concrete will be assumed to act through the bottom flange plate.

Unfactored loads to steel beam.

Loads acting:

PC units	=	4 × 8 × 8/2	= 128.0 kN	· eccentric to UC
Steelwork	=	0.2 × 8 × 8	= 12.8 kN	· on CL to UC
Conc. encasement	=	0.3 × 8 × 8	= 19.2 kN	· on CL to UC
Concrete slab	=	3.2 × 8 × 8/2	= 102.4 kN	· eccentric to UC

Unfactored loads applied to composite section

C & S	=	0.4 × 8 × 8/2	= 12.8 kN	· eccentric to UC
Imposed	=	5.6 × 8 × 8/2	= 179.2 kN	· eccentric to UC

Loads to steel beam

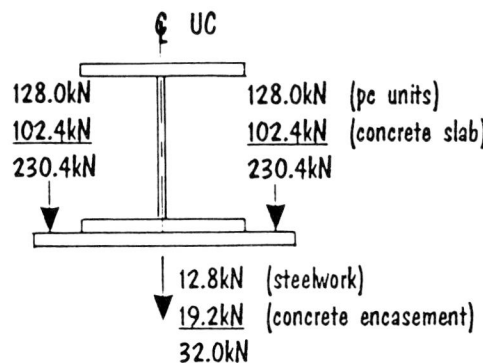

	Job No.	PUB 810	Sheet 46 of 56	Rev.
The Steel Construction Institute Silwood Park Ascot Berks SL5 7QN Telephone: (0344) 23345 Fax: (0344) 22944 **CALCULATION SHEET**	Job Title	Worked Example No 3		
	Subject	Serviceability Stresses to Steel Beam		
	Client	Made by DLM	Date Oct 1991	
		Checked by JWR	Date Oct 1991	

Loads applied to composite section

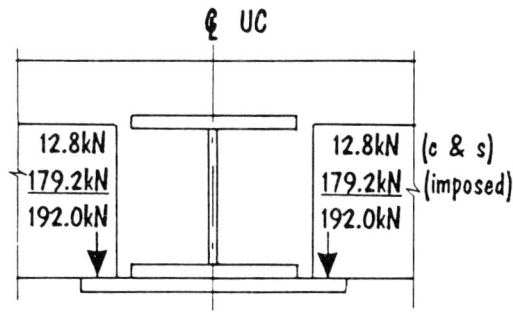

Stresses due to overall bending of the section.

Steel beam

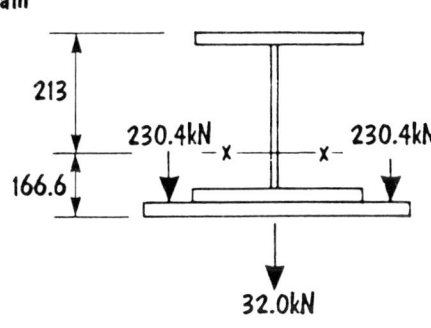
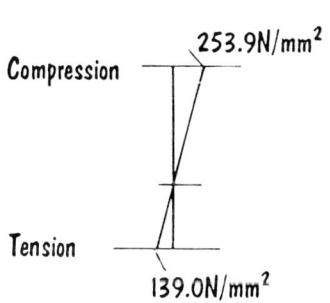

I_{xx} = 41347 cm⁴

Z_c = 41347/21.3 = 1941 cm³

Z_t = 41347/11.66 = 3546 cm³

Max. moment = (230.4 + 230.4 + 32.0) 8/8 = 492.8 kN.m

$$\sigma_c = \frac{492.8 \times 10^3}{1941} = 253.9 \text{ N/mm}^2 \text{ at mid span}$$

$$\sigma_t = \frac{492.8 \times 10^3}{3546} = 139.0 \text{ N/mm}^2 \text{ at mid span}$$

	Job No.	PUB 810	Sheet **47** of **56**	Rev.
The Steel Construction Institute	Job Title	Worked Example No 3		
Silwood Park Ascot Berks SL5 7QN Telephone: (0344) 23345 Fax: (0344) 22944	Subject	Serviceability Stresses to Composite Section		
CALCULATION SHEET	Client	Made by DLM	Date Oct 1991	
		Checked by JWR	Date Oct 1991	

Composite section

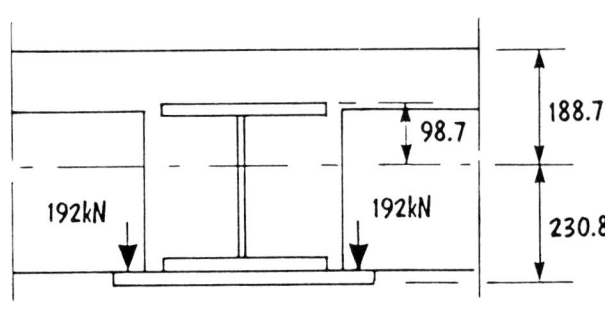
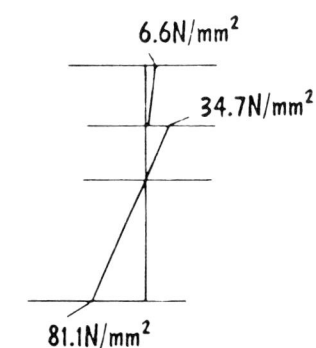

I_{xx} = 109274 cm^4

Max moment = (192 + 192) × 8/8 = 384.0 kN.m

Z (concrete) = $\dfrac{109274 \times 10}{18.87}$ = 57909 cm^3

Z_c = $\dfrac{109274}{9.87}$ = 11071 cm^3

Z_t = $\dfrac{109274}{23.08}$ = 4735 cm^3

Concrete stress = $\dfrac{384 \times 10^3}{57909}$ = 6.6 N/mm^2 < $\dfrac{30}{2}$ = 15 N/mm^2 OK

σ_c = $\dfrac{384 \times 10^3}{11071}$ = 34.7 N/mm^2

σ_t = $\dfrac{384 \times 10^3}{4735}$ = 81.1 N/mm^2

	Job No.	PUB 810	Sheet 48 of 56	Rev.
The Steel Construction Institute Silwood Park Ascot Berks SL5 7QN Telephone: (0344) 23345 Fax: (0344) 22944 **CALCULATION SHEET**	Job Title	Worked Example No 3		
	Subject	Combined Stresses		
	Client	Made by DLM	Date Oct 1991	
		Checked by JWR	Date Oct 1991	

Combined longitudinal stresses for the steel beam.

$\sigma_c = 253.9 + 34.7 = 288.6 \text{ N/mm}^2 < 345 \text{ N/mm}^2$ OK

$\sigma_t = 139.0 + 81.1 = 220.1 \text{ N/mm}^2 <^* 355 \text{ N/mm}^2$ OK

* plate stress check.

The extreme fibre tensile stresses have to be combined with the transverse bending stresses.

$Z_e \text{ (plate modulus)} = \dfrac{1000 \times 15^2}{6 \times 10^3} = 37.5 \text{ cm}^3 \text{ per m.}$

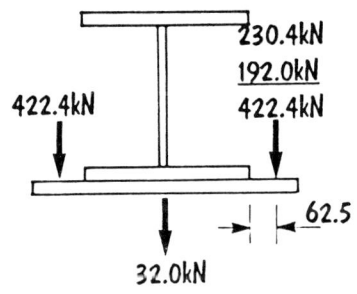

Plate BM $= \dfrac{422.4}{8} \times \dfrac{62.5}{10^3}$

$= 3.3 \text{ kN.m per m}$

$\sigma_{tt} = \dfrac{3.3 \times 10^3}{37.5} = 88 \text{ N/mm}^2$

Combined stress must be less than or equal to p_y

$(\sigma_{tt}^2 + \sigma_t^2 + \sigma_{tt}\sigma_t)^{1/2} \not> p_y$

$= (88^2 + 220.1^2 + 88 \times 220.1)^{1/2}$

$= 274.9 \text{ N/mm}^2 < 355 \text{ N/mm}^2$ OK

The maximum serviceability stress occured in the compression flange of the UC.

∴ <u>Serviceability stresses satisfactory</u>

	Job No.	PUB 810	Sheet 49 of 56	Rev.
The Steel Construction Institute	Job Title	Worked Example No 3		
Silwood Park Ascot Berks SL5 7QN Telephone: (0344) 23345 Fax: (0344) 22944	Subject	Transverse Reinforcement		
CALCULATION SHEET	Client	Made by DLM	Date Oct 1991	
		Checked by JWR	Date Oct 1991	

TRANSVERSE REINFORCEMENT

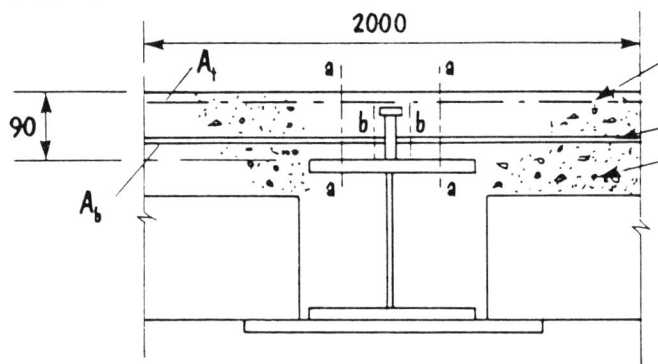

A142 mesh
16⌀ bars @ 540 crs.
Grade 30 NW concrete
19⌀ studs x 70 (LAW) @ 180mm crs.

Mesh Reinforcement
$f_y = 460 N/mm^2$

$A_b = 201/0.54 = 372.2 mm^2/m$

Note: Where possible keep the bar spacing in multiples of the stud spacing
i.e. 3 × 180 = 540 mm

Shear force, v, to be resisted per unit length

$$v = \frac{NQ_p}{s} = \frac{1 \times 69.6}{0.18} = 386.7 \text{ kN/m}$$

Two potential shear failure surfaces to be considered.

Length of shear surfaces:

Condition (i) (a-a) = 90 mm (for one surface only)
Condition (ii) (b-b) = 2h + 31 = 2 × 70 + 31 = 171 mm (Total)

Note: h = height of shear stud connector
31 mm is the diameter of the shear stud head.

Condition (i) Shear surface (a-a)

Resistance of concrete flange, v_r

$v_r = 0.7 A_{sv}f_y + 0.03\eta A_{cv}f_{cu}$ but $\leq 0.8\eta A_{cv}\sqrt{f_{cu}}$

Where:

$A_{sv} = A_t + A_b = 142 + 372.2 = 514.2 \text{ mm}^2/m$

$A_{cv} = 90 \times 10^3 \text{ mm}^2/m$

$f_y = 460 \text{ N/mm}^2$

	Job No.	PUB 810	Sheet 50 of 56	Rev.
The Steel Construction Institute	Job Title	Worked Example No 3		
Silwood Park Ascot Berks SL5 7QN Telephone: (0344) 23345 Fax: (0344) 22944	Subject	Transverse Reinforcement cont'd		
CALCULATION SHEET	Client	Made by DLM	Date Oct 1991	
		Checked by JWR	Date Oct 1991	

f_{cu} = 30 N/mm²
η = 1.0 for NW Concrete

$$v_r = \frac{0.7 \times 514.2 \times 460}{10^3} + \frac{0.03 \times 1.0 \times 90 \times 10^3 \times 30}{10^3}$$

$\quad$ = 165.6 + 81.0

$\quad$ = 246.6 kN/m > 386.7/2 = 193.4 kN/m OK

Check condition

$$v_r = \frac{0.8 \times 1.0 \times 90 \times 10^3 \times (30)^{1/2}}{10^3}$$

$\quad$ = 394.4 kN/m > 246.6 kN/m OK

∴ <u>first condition satisfactory</u>

Condition (ii) Shear surface (b-b)

A_{sv} = $2A_b$ = 2 × 372.2 = 744.4 mm²/m

A_{cv} = 171 × 10³ mm²/m

$$v_r = \frac{0.7 \times 744.4 \times 460}{10^3} + \frac{0.03 \times 1.0 \times 171 \times 10^3 \times 30}{10^3}$$

$\quad$ = 239.7 + 153.9

$\quad$ = 393.6 kN/m > 386.7 kN/m OK

Check condition

$$v_r = \frac{0.81 \times 1.0 \times 171 \times 10^3 \times \sqrt{30}}{10^3}$$

$\quad$ = 749.3 kN/m > 393.6 kN/m OK

∴ Second condition satisfactory

∴ <u>Use A142 Mesh and 16∅ bars at 540 crs</u>

	Job No.	PUB 810		Sheet 51 of 56	Rev.
The Steel Construction Institute Silwood Park Ascot Berks SL5 7QN Telephone: (0344) 23345 Fax: (0344) 22944	Job Title	Worked Example No 3			
	Subject	Natural Frequency			
CALCULATION SHEET	Client		Made by DLM	Date Oct 1991	
			Checked by JWR	Date Oct 1991	

NATURAL FREQUENCY

The loading = Dead + 10% Imposed (not including partitions)

Dead = $128 \times 2 + 12.8 + 19.2 + 102.4 \times 2 + 25.6$
 = 518.4 kN

Imposed = 5.0 kN/m²

Imposed load = 0.5×64 = 32 kN
Total load = $518.4 + 32$ = 550.4 kN

$$\delta_{SW} = \frac{5 \times 550.4 \times 8000^3}{384 \times 205 \times 109274 \times 10^4}$$

= 16.3 mm

Frequency = $\dfrac{18}{\sqrt{16.3}}$

= 4.5 H_z > 4.0 OK

∴ Floor structure satisfactory for natural frequency

If this check had not satisfied the 4.0 H_z limit, then Reference (9) dealing with vibration provides a more detailed design procedure.

	Job No.	PUB 810		Sheet 52 of 56	Rev.
The Steel Construction Institute	Job Title	Worked Example No 3			
Silwood Park Ascot Berks SL5 7QN Telephone: (0344) 23345 Fax: (0344) 22944	Subject	Beam to Column Connection			
CALCULATION SHEET	Client		Made by DLM		Date Oct 1991
			Checked by JWR		Date Oct 1991

"SLIMFLOR" BEAM TO COLUMN CONNECTION

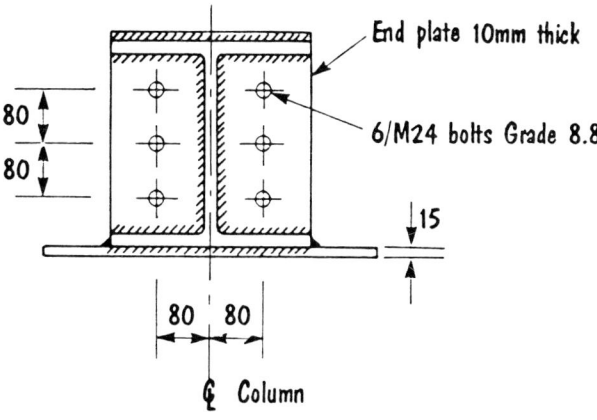

Note: For bolt cross-sections and end plate thickness see Section 4 of this publication.

Factored loads

Max. Vertical Shear
= 649.6 kN

Check using 6-M24 Bolts Grade 8.8

Shear per bolt
= $\dfrac{649.6}{6}$ = 108.3 kN < 132.0 kN (shear capacity M24) OK

Plate bearing (Grade 50)
= $24 \times 10 \times \dfrac{550}{10^3}$ = 132.0 kN > 108.3 kN OK

Plate bearing (Grade 43)
= $24 \times 10 \times \dfrac{460}{10^3}$ = 110.0 kN > 108.3 kN OK

∴ Use Grade 43 steel for the end plate.

Note: The SCI publication "Guide to BS 5950 Volume 1" (Reference 13) gives values of bolt capacities.

	Job No.	PUB 810		Sheet 53 of 56	Rev.
The Steel Construction Institute Silwood Park Ascot Berks SL5 7QN Telephone: (0344) 23345 Fax: (0344) 22944	Job Title	Worked Example No 3			
	Subject	Beam to Column Connection cont'd			
CALCULATION SHEET	Client		Made by DLM	Date Oct 1991	
			Checked by JWR	Date Oct 1991	

Check on bolt forces due to out of balance loads in the construction stage.

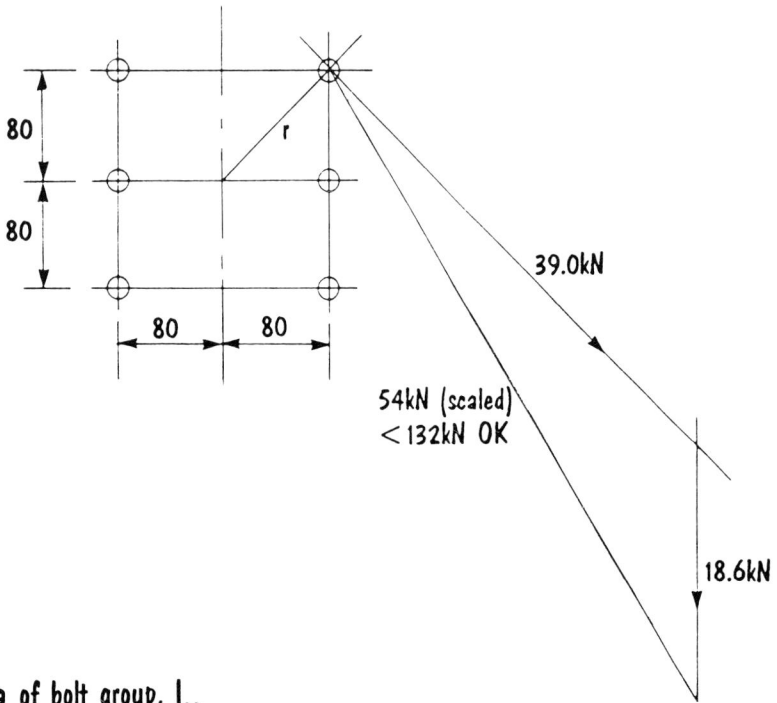

Polar Inertia of bolt group, I_{oo}

I_{oo} = $(4 \times 8^2) + (6 \times 8^2)$ = 640 cm^4

distance, r = $8\sqrt{2}$ = 11.3 cm

Modulus = 640/11.3 = 56.6 cm^3

Assumed Construction stage loading

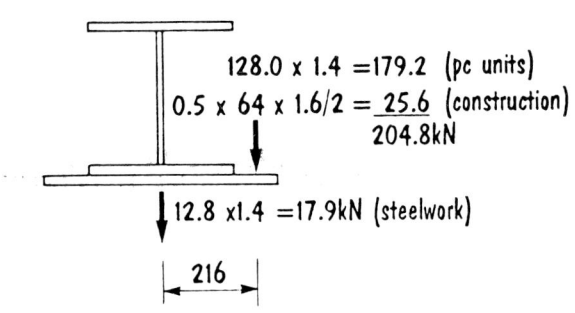

128.0 x 1.4 = 179.2 (pc units)
0.5 x 64 x 1.6/2 = 25.6 (construction)
204.8 kN

12.8 x 1.4 = 17.9 kN (steelwork)

216

	Job No.	PUB 810	Sheet 54 of 56	Rev.
The Steel Construction Institute Silwood Park Ascot Berks SL5 7QN Telephone: (0344) 23345 Fax: (0344) 22944	Job Title	Worked Example No 3		
	Subject	Beam to Column Connection cont'd		
CALCULATION SHEET	Client	Made by DLM	Date Oct 1991	
		Checked by JWR	Date Oct 1991	

Torsional moment per connection

$$= \frac{204.8}{2} \times \frac{216}{10^3} = 22.1 \text{ kN.m}$$

Vertical shear per connection

$$= (204.8 + 17.9)/2 = 111.4 \text{ kN} \qquad \text{per bolt} = \frac{111.4}{6} = 18.6 \text{ kN}$$

Component force due to torsion

$$= \frac{22.1 \times 10^2}{56.6} = 39.0 \text{ kN}$$

∴ Bolts satisfactory for max shear and construction stage out of balance loads - Use 6/M24 Grade 8.8 bolts

For the weld design all of the vertical shear has been assumed to act through the flange plate except for the own weight of steelwork and concrete encasement.

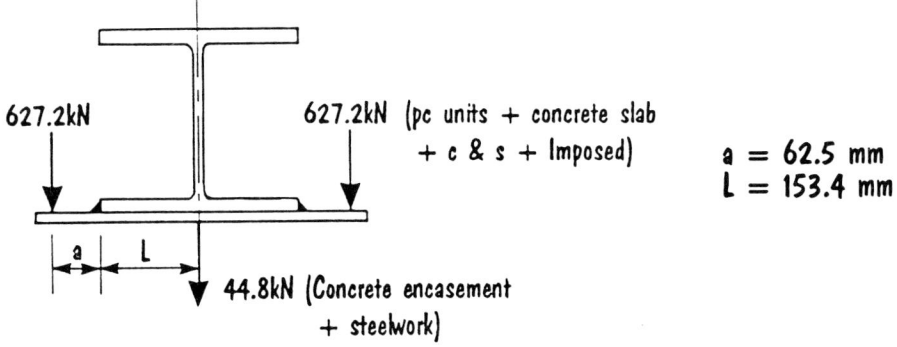

$a = 62.5$ mm
$L = 153.4$ mm

Under balanced load conditions the curvature at the centre of the beam is zero. The assumption is that the plate then acts as a propped cantilever.

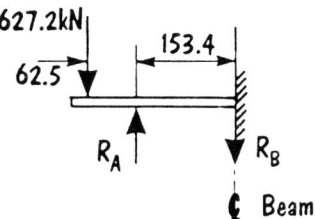

SHEAR FORCE AT WELD POSITION, R_A

$$R_A = 627.2 \left(1 + \frac{3 \times 62.5}{2 \times 153.4}\right) = 1010.5 \text{ kN} \quad > 1.4P = 878 \text{ kN OK}$$

Load/m = 1010.5/8 = 126.3 kN/m

Longitudinal force, R_P

R_P = Max. possible force in plate = 2524.5 kN

Note: Using the analogy of prying forces in bolted connections, it can be shown that when (L) is much greater than (a) the prying force will act closer to the weld than (2/3)L. Hence, the magnitude will be larger than that given by the simple expression $P(1 + \frac{1.5a}{L})$.

To avoid this from happening a limit of 1.4P has been adopted.

The Steel Construction Institute	Job No.	PUB 810		Sheet 56 of 56	Rev.
Silwood Park Ascot Berks SL5 7QN Telephone: (0344) 23345 Fax: (0344) 22944	Job Title	Worked Example No 3			
	Subject	Weld Design cont'd			
CALCULATION SHEET	Client		Made by DLM	Date Oct 1991	
			Checked by JWR	Date Oct 1991	

Assume an elastic stress distribution for the transfer of longitudinal shear between the UC and the plate.

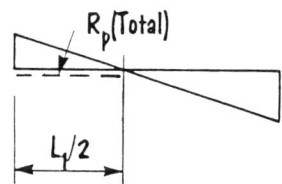

Hence max shear/m, s is given by:

$$\frac{1}{2} \times S \times \frac{L_1}{2} = R_p$$

$$\therefore \text{shear per weld} = \frac{4R_p}{L_1} \times \frac{1}{2} = \frac{2R_p}{L_1}$$

$$= \frac{2 \times 2524.5}{8}$$

$$= 631.1 \text{ kN/m}$$

HORIZONTAL SHEAR FORCE (Assume 8mm fillet)

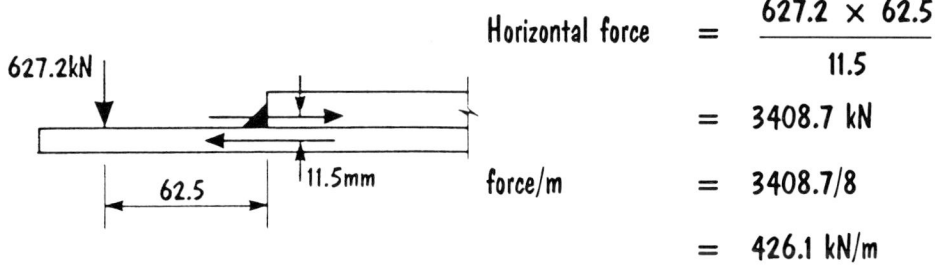

Horizontal force $= \dfrac{627.2 \times 62.5}{11.5}$

$= 3408.7$ kN

force/m $= 3408.7/8$

$= 426.1$ kN/m

Resultant Force $= (0.126^2 + 0.631^2 + 0.426^2)^{1/2} = 0.8$ kN/mm

Min. Weld Size $=$ 8mm fillet

Capacity $= 8 \times 0.7 \times 255/10^3 = 1.428$ kN/mm

$\therefore$ 1.428 > 0.8 kN/mm OK Use 8mm Fillet Weld

APPENDIX B: Formulae for elastic section properties and plastic moment capacity

(i) Steel section only
(ii) Steel section plus concrete

(i) **STEEL SECTION ONLY**

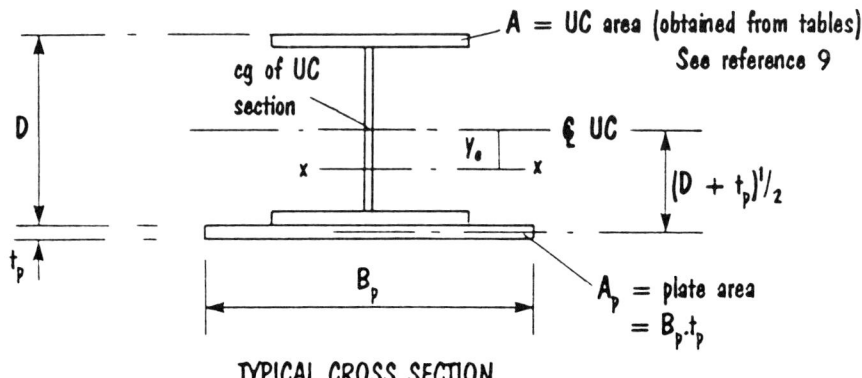

TYPICAL CROSS SECTION

Position of NA, y_e
Moments about cg. of UC section

$$(D + t_p) A_p/2 = (A_p + A) y_e$$

$$\therefore y_e = \frac{(D + t_p) A_p}{2 A_T} \quad \text{where } A_T = A_p + A$$

I_{xx}, second moment of area of plated steel section

$$I_{xx} = (I_{x(uc)} + A y_e^2) + A_p \left(\frac{D}{2} + \frac{t_p}{2} - y_e\right)^2$$

$$I_y = I_{y(uc)} + t_p \frac{B_p^3}{12}$$

SECTION MODULI, Z_x

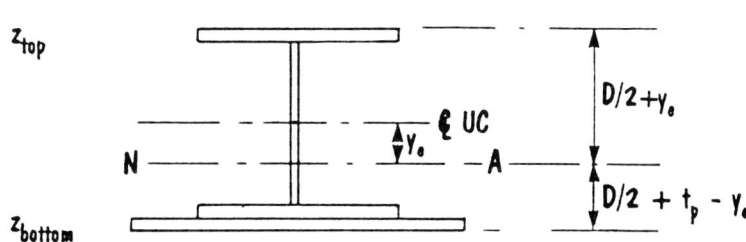

$$Z_{top} = \frac{I_{xx}}{(D/2 + y_e)} \quad \text{(Compression)}$$

$$Z_{bottom} = \frac{I_{xx}}{(D/2 + t_p - y_e)} \quad \text{(Tension)}$$

(ii) **STEEL SECTION PLUS CONCRETE**

a) Uncracked properties

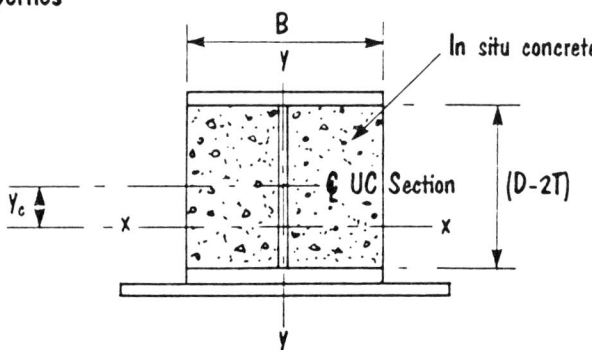

modular ratio, α_e

Concrete Area, A_c (Steel units)

$$A_c = \frac{B(D-2T)}{\alpha_e}$$

Concrete second moment of area, I_{cx}

$$I_{cx} = \frac{B(D-2T)^3}{12\alpha_e} \qquad I_{cy} = \frac{(D-2T)B^3}{12\alpha_e}$$

$$I_x = (I_{x(uc)} + Ay_c^2) + A_p\left(\frac{D+t_p}{2} - y_c\right)^2 + B\frac{(D-2T)^3}{12\alpha_e} + A_c y_c^2$$

$$I_y = I_{y(uc)} + \frac{t_p(B_p)^3}{12} + \frac{(D-2T)B^3}{12\alpha_e}$$

where:

$$\left(\frac{D}{2} + \frac{t_p}{2}\right) A_p = (A_T + A_c) y_c$$

$$\therefore y_c = \frac{A_p(D+t_p)}{2(A_T + A_c)}$$

b) Cracked Properties

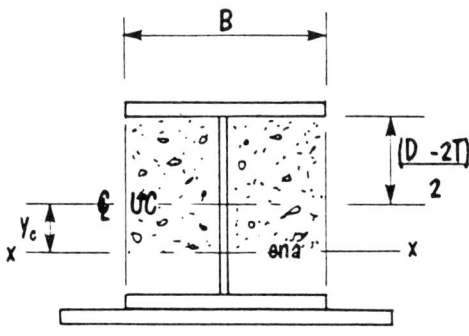

Assume n. a. is below UC Centre Line (CL) but if it is actually above then same formulae apply but y_c will have a negative value.

Modular ratio, α_e

Concrete area, $A_c = \dfrac{B}{\alpha_e}\left[\left(\dfrac{D-2T}{2}\right) + y_c\right]$ (in steel units)

1st moment of area about UC CL to find y_c

$$(A_T + A_c)\, y_c = -\dfrac{A_c}{2}\left[\left(\dfrac{D-2T}{2}\right) - y_c\right] + A_p\left(\dfrac{D}{2} + \dfrac{t_p}{2}\right)$$

$$\left\{A_T + \dfrac{B}{\alpha_e}\left[\left(\dfrac{D-2T}{2}\right)+y_c\right]\right\} y_c = \dfrac{-B}{2\alpha_e}\left[\left(\dfrac{D-2T}{2}\right)+y_c\right]\left[\left(\dfrac{D-2T}{2}\right)-y_c\right] + A_p\left(\dfrac{D}{2}+\dfrac{t_p}{2}\right)$$

$$= \dfrac{-B}{2\alpha_e}\left[\left(\dfrac{D-2T}{2}\right)^2 - y_c^2\right] + \dfrac{A_p}{2}(D+t_p)$$

The Steel Construction Institute		Job No.	PUB 810	Sheet	5 of 22	Rev.
Silwood Park Ascot Berks SL5 7QN Telephone: (0344) 23345 Fax: (0344) 22944		Job Title	Appendix B			
		Subject	Elastic Section Properties - Slimflor beam			
CALCULATION SHEET		Client		Made by DLM		Date Oct 1991
				Checked by JWR		Date Oct 1991

$$\therefore y_c^2\left(\frac{B}{\alpha_e}-\frac{B}{2\alpha_e}\right)+y_c\left\{A_T+\frac{B}{\alpha_e}\left(\frac{D-2T}{2}\right)\right\}+\left\{\frac{B}{2\alpha_e}\left(\frac{D-2T}{2}\right)^2-A_p\left(\frac{D}{2}+\frac{t_p}{2}\right)\right\}=0$$

Hence $y_c^2\,(k_1) + y_c\,(k_2) + (k_3) = 0$

where: $\quad k_1 \;=\; \dfrac{B}{2\alpha_e}$

$\quad\quad\quad\;\; k_2 \;=\; \left\{A_T + \dfrac{B}{\alpha_e}\left(\dfrac{D-2T}{2}\right)\right\}$

$\quad\quad\quad\;\; k_3 \;=\; \left\{\dfrac{B}{2\alpha_e}\left(\dfrac{D-2T}{2}\right)^2 - A_p\left(\dfrac{D+t_p}{2}\right)\right\}$

and $\therefore y_c = \dfrac{-k_2 \pm \sqrt{k_2^2 - 4k_1 k_3}}{2k_1}$

$$I_x = (I_{uc} + Ay_c^2) + A_p\left(\frac{D+t_p}{2}-y_c\right)^2 + \frac{B}{12\,\alpha_e}\left[\left(\frac{D-2T}{2}\right)+y_c\right]^3 + A_c\left[\left(\frac{D-2T}{4}\right)-\frac{y_c}{2}\right]^2$$

$\quad = (k_4) + (k_5) + (k_6) + (k_7)$

Z (concrete) $= \dfrac{I_x \alpha_e}{\left(\dfrac{D-2T}{2}+y_c\right)}$ $\quad$ Z (steel top) $= \dfrac{I_x}{\left(\dfrac{D}{2}+y_c\right)}$ (compression)

Z (steel bottom) $= \dfrac{I_x}{\left(\dfrac{D}{2}+t_p-y_c\right)}$ (tension)

ELASTIC NEUTRAL AXIS IN CONCRETE SLAB - ZONE 1

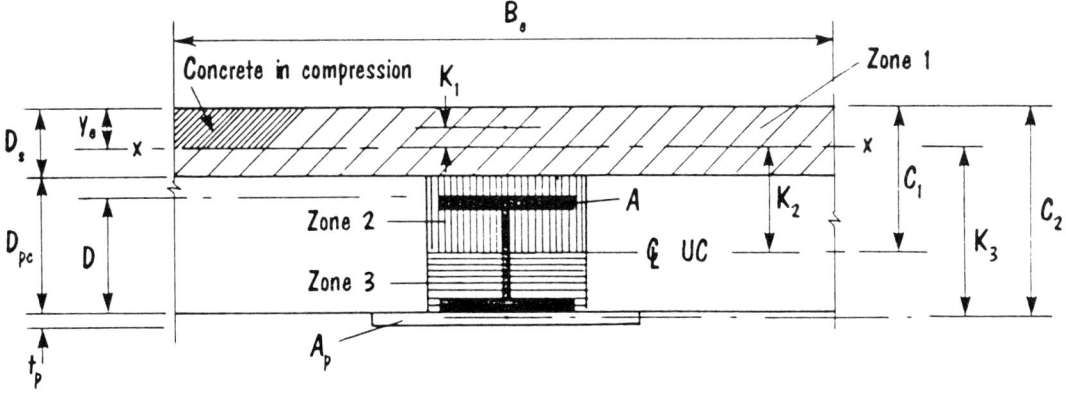

Moments taken about the top of the concrete slab.

Solve the following quadratic equation for y_e

$$\frac{B_e y_e}{\alpha_e} \cdot \frac{y_e}{2} + A(c_1) + A_p(c_2) = \left(\frac{B_e y_e}{\alpha_e} + A + A_p\right) y_e$$

$$\frac{B_e y_e^2}{2\alpha_e} + A(c_1) + A_p(c_2) = \frac{B_e y_e^2}{\alpha_e} + A y_e + A_p y_e$$

$$-\frac{B_e y_e^2}{2\alpha_e} + A(c_1) + A_p(c_2) - A y_e - A_p y_e = 0$$

$$\frac{B_e y_e^2}{2\alpha_e} + y_e(A + A_p) - A(c_1) - A_p(c_2) = 0$$

where:

$c_1 = D_s + D_{pc} - \dfrac{D}{2}$

$c_2 = D_s + D_{pc} + t_p/2$

Notes:
1. Cracked Inertia/Section moduli is used for strength when the ENA is in the concrete.
2. If $D_{pc} < D$, replace D_{pc} with D in the above equations.

ELASTIC NEUTRAL AXIS IN CONCRETE SLAB - ZONE 1 (Continued)

Cracked Inertia, I_{xx}

Concrete Inertia, I_{xc}

$$I_{xc} = \frac{B_e y_e^3}{3\alpha_e} \qquad (k_1 = y_e/2)$$

$$I_{xx} = \frac{B_e y_e^3}{3\alpha_e} + I_{x(uc)} + A(k_2)^2 + A_p(k_3)^2$$

where:

$k_2 = D_s + D_{pc} - D/2 - y_e$
$k_3 = D_s + D_{pc} + t_p/2 - y_e$

Section Moduli

$$Z_x \text{ (Concrete)} = \frac{I_{xx} \alpha_e}{y_e}$$

$$Z_x \text{ (Steel)} = \frac{I_{xx}}{D_s + D_{pc} + t_p - y_e}$$

ELASTIC NEUTRAL AXIS IN ZONE 2 AND 3

Zone 2

Other dimensions as Zone 1 diagram.

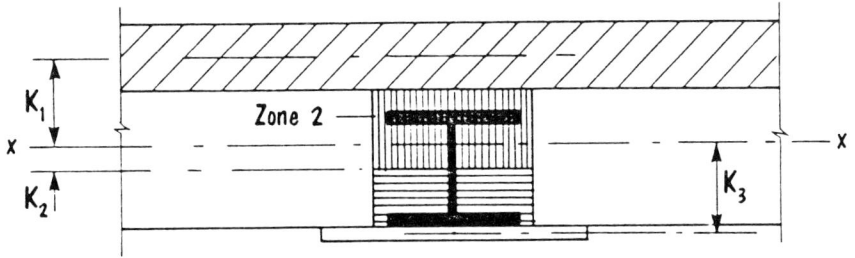

General expression for y_e

$$y_e = \frac{\dfrac{B_e}{2\alpha_e} \cdot D_S^2 + A(c_1) + A_p(c_2)}{A + A_p + \dfrac{B_e}{\alpha_e} D_S}$$

where:

$c_1 = D_S + D_{pc} - D/2 \qquad c_2 = D_S + D_{pc} + t_p/2$

Inertia, I_{xx}

$$I_{xx} = \frac{B_e}{\alpha_e} D_S \left(\frac{D_S^2}{12} + k_1^2 \right) + I_{x(UC)} + A k_2^2 + A_p k_3^2$$

where:

$k_1 = y_e - D_S/2$ $\qquad Z_x \text{(Concrete)} = \dfrac{I_x \alpha_e}{y_e}$

$k_2 = D_S + D_{pc} - D/2 - y_e$ $\qquad Z_x \text{(Steel)} = \dfrac{I_x}{D_S + D_{pc} + t_p - y_e}$

$k_3 = D_S + D_{pc} + t_p/2 - y_e$

Elastic Neutral Axis in Zone 3

The expression for Inertia and Section moduli are as above but the k factors become:

$k_1 = y_e - D_S/2$
$k_2 = y_e - D_S - D_{pc} + D/2$
$k_3 = D_S + D_{pc} + t_p/2 - y_e$

The Steel Construction Institute	Job No. **PUB 810**		Sheet **10** of **22**	Rev.
Silwood Park Ascot Berks SL5 7QN Telephone: (0344) 23345 Fax: (0344) 22944	Job Title **Appendix B**			
	Subject **Elastic Section Properties - Composite Section**			
CALCULATION SHEET	Client	Made by **DLM**	Date **Oct 1991**	
		Checked by **JWR**	Date **Oct 1991**	

If the ENA is in Zone 1 then:

a) for the cracked section properties - solve the following quadratic for y_e

$$\frac{B_e y_e^2}{2\alpha_e} + y_e (A + A_p) - A(c_1) - A_p(c_2) = 0$$

$$I_{xx} = \frac{B_e y_e^3}{3\alpha_e} + I_{x(uc)} + A(k_2)^2 + A_p(k_3)^2$$

$$Z_x \text{(concrete)} = \frac{I_{xx} \cdot \alpha_e}{y_e}, \quad Z_x \text{(steel)} = \frac{I_{xx}}{D_s + D_{pc} + t_p - y_e}$$

where:
$c_1 = D_s + D_{pc} - D/2$
$c_2 = D_s + D_{pc} + t_p/2$
$k_2 = D_s + D_{pc} - D/2 - y_e$
$k_3 = D_s + D_{pc} + t_p/2 - y_e$

b) for the uncracked section properties -

$$I_{xx} = \frac{B_e D_s}{\alpha_e}\left(\frac{D_s^2}{12} + k_1^2\right) + I_{x(uc)} + A k_2^2 + A_p k_3^2$$

$$Z_x \text{(concrete)} = \frac{I_{xx} \alpha_e}{y_e}, \quad Z_x \text{(steel)} = \frac{I_{xx}}{D_s + D_{pc} + t_p - y_e}$$

	Job No.	PUB 810	Sheet 11 of 22	Rev.
The Steel Construction Institute Silwood Park Ascot Berks SL5 7QN Telephone: (0344) 23345 Fax: (0344) 22944	Job Title	**Appendix B**		
	Subject	**Elastic Section Properties – Composite Section**		
CALCULATION SHEET	Client	Made by DLM	Date Oct 1991	
		Checked by JWR	Date Oct 1991	

where:
$$y_c = \frac{\dfrac{B_e D_s^2}{2\alpha_e} + A(c_1) + A_p(c_2)}{A + A_p + \dfrac{B_e}{\alpha_e} \cdot D_s}$$

c_1 and c_2 as above

$k_1 = y_e - D_s/2$
$k_2 = D_s + D_{pc} - D/2 - y_e$
$k_3 = D_s + D_{pc} + t_p/2 - y_e$

If the ENA is in zone 2 then:

I_{xx} as for zone 1 (b)

Z_{xx} as for zone 1 (b)

where:
$k_1 = y_e - D_s/2$
$k_2 = D_s + D_{pc} - D/2 - y_e$
$k_3 = D_s + D_{pc} + t_p/2 - y_e$

If the ENA is in zone 3 then:

I_{xx} as for zone 1 (b)

Z_{xx} as for zone 1 (b)

where:
$k_1 = y_e - D_s/2$
$k_2 = y_e - D_s - D_{pc} + D/2$
$k_3 = D_s + D_{pc} + t_p/2 - y_e$

Notes:
1) If $D_{pc} < D$, replace D_{pc} with D in the above equations
2) When the ENA is in Zone 1, the uncracked properties may be used for the deflection calculations but the cracked properties should be used for stress calculations.

	Job No.	PUB 810	Sheet 12 of 22	Rev.
The Steel Construction Institute	Job Title	Appendix B		
Silwood Park Ascot Berks SL5 7QN Telephone: (0344) 23345	Subject	Plastic Moment Capacity - Slimflor beam		
Fax: (0344) 22944	Client	Made by DLM	Date Oct 1991	
CALCULATION SHEET		Checked by JWR	Date Oct 1991	

PLASTIC MOMENT CAPACITY

Resistances

The plastic moment capacity is expressed in terms of the resistance of various elements of the beam as follows.

Resistance of concrete flange, $R_c = 0.45 f_{cu} B_e (D_s - D_{pc})$

Resistance of steel flange, $R_f = BT p_y$

Resistance of flange plate, $R_p = B_p t_p p_y$

Resistance of shear connection, $R_q = NQ$

Resistance of steel beam, $R_s = A p_y$

Resistance of clear web depth, $R_v = d t_w p_y$

Resistance of overall web depth, $R_w = R_s - 2R_f$

where:

A is the area of the steel beam;

A_p is the area of the flange plate;

B is the breadth of the steel flange;

B_e is the effective breadth of the concrete flange;

D_{pc} is the depth of the pc unit;

D_S is the depth of the insitu concrete

d is the clear depth of the web;

f_{cu} is the characteristic strength of the concrete;

T is the flange thickness

t_w is the web thickness

t_p is the bottom flange plate thickness

S_x plastic section moduli of the UC section

N is the actual number of shear connectors for positive moments as relevant (minimum number, one side of the point of maximum moment);

	Job No.	PUB 810	Sheet 13 of 22	Rev.
The Steel Construction Institute	Job Title	Appendix B		
Silwood Park Ascot Berks SL5 7QN Telephone: (0344) 23345 Fax: (0344) 22944	Subject	Plastic Moment Capacity - Slimflor beam		
CALCULATION SHEET	Client	Made by DLM	Date Oct 1991	
		Checked by JWR	Date Oct 1991	

Case 1 SP (Steel Plastic): pna in web

Where $R_w > R_p$

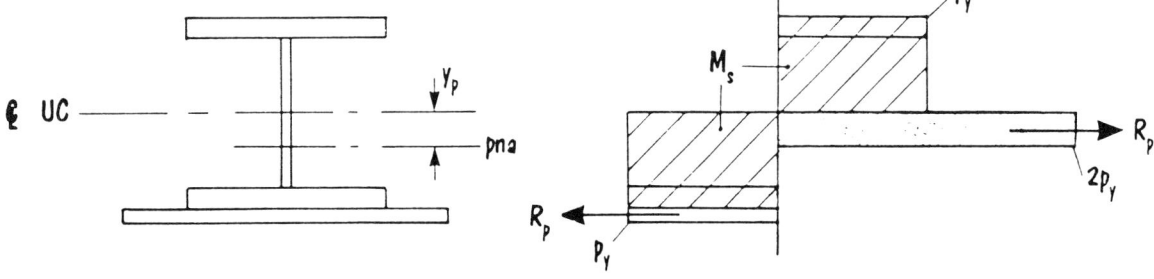

$$y_p = \frac{R_p}{2t_w p_y}$$

Moments about Centre Line of Beam

$$M_c = M_S + R_p\left(\frac{D}{2} + \frac{t_p}{2}\right) - R_p \cdot y_p/2$$

$$= M_S + \frac{R_p}{2}(D + t_p) - \frac{R_p^2}{4p_y t_w}$$

$$M_S = S_x p_y$$

$$\boxed{M_c = M_S + \frac{R_p}{2}(D + t_p) - \frac{R_p^2}{4p_y t_w}}$$

$$y_p = \frac{R_p}{2t_w p_y}$$

Case 2 SP (Steel Plastic): pna in the UC flange

Where $(R_p \geq R_w)$ & $(R_s > R_p)$

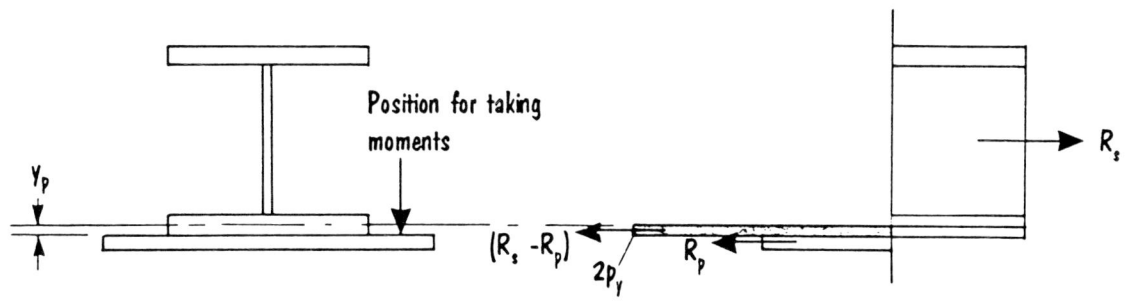

$$M_c = R_s.D/2 + R_p.t_p/2 - (R_s - R_p)\, y_p/2$$

where: $\quad y_p = \dfrac{(R_s - R_p)}{2.B.p_y}$

$$\therefore M_c = R_s.D/2 + R_p.t_p/2 - \dfrac{(R_s - R_p)^2}{4.B.p_y}$$

$$\boxed{M_c = R_s \dfrac{D}{2} + R_p \cdot \dfrac{t_p}{2} - \dfrac{(R_s - R_p)^2}{4.B.p_y}}$$

$$y = \left(\dfrac{R_s - R_p}{2.B.p_y}\right)$$

The Steel Construction Institute	Job No.	PUB 810	Sheet 15 of 22	Rev.
Silwood Park Ascot Berks SL5 7QN Telephone: (0344) 23345 Fax: (0344) 22944	Job Title	Appendix B		
	Subject	Plastic Moment Capacity - Slimflor beam		
CALCULATION SHEET	Client		Made by DLM	Date Oct 1991
			Checked by JWR	Date Oct 1991

Case 3 SP(Steel Plastic) ~ pna in the bottom flange plate.

$R_p > R_w$ & $R_p > R_s$

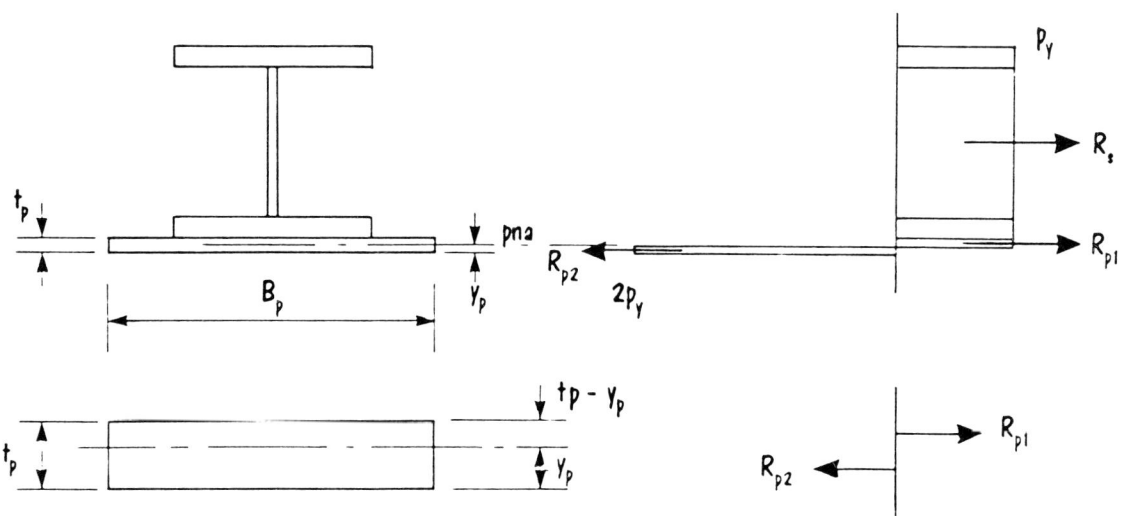

$R_{p2} = R_s + R_{p1}$ basic eqn.

$B_p \cdot p_y \cdot y_p = R_s + B_p \cdot p_y (t_p - y_p)$

$B_p \cdot p_y \cdot y_p = R_s + B_p \cdot p_y \cdot t_p - B_p \cdot p_y \cdot y_p$

$2 B_p \cdot p_y \cdot y_p = R_s + B_p \cdot p_y \cdot t_p = R_s + R_p$

$\therefore y_p = \dfrac{R_s + R_p}{2 \cdot B_p \cdot p_y}$

$R_{p2} = B_p \cdot p_y \cdot y_p$

$R_{p1} = (t_p - y_p) B_p \cdot p_y$

Also -

$R_{p2} = R_p \cdot y_p / t_p$

$R_{p1} = \left(\dfrac{t_p - y_p}{t_p} \right) R_p$

$R_p = R_{p1} + R_{p2}$

$$M_c = R_{p2}\, y_p/2 + \frac{R_{p1}}{2}(t_p - y_p) + R_s\left(\frac{D}{2} + t_p - y_p\right)$$

$$= \left(\frac{R_p \cdot y_p}{t_p}\right)\frac{y_p}{2} + \frac{R_p}{2}\left(\frac{t_p - y_p}{t_p}\right)(t_p - y_p) + R_s\left(\frac{D}{2} + t_p - y_p\right)$$

$$= \frac{R_p}{2t_p}\left[y_p^2 + (t_p - y_p)^2\right] + R_s(D/2 + t_p - y_p)$$

$$\boxed{M_c = \frac{R_p}{2t_p}\left[y_p^2 + (t_p - y_p)^2\right] + R_s(D/2 + t_p - y_p)}$$

$$y_p = \frac{R_s + R_p}{2.B_p.p_y}$$

CASE 1 FS (Full Shear) - pna lies within the UC top flange

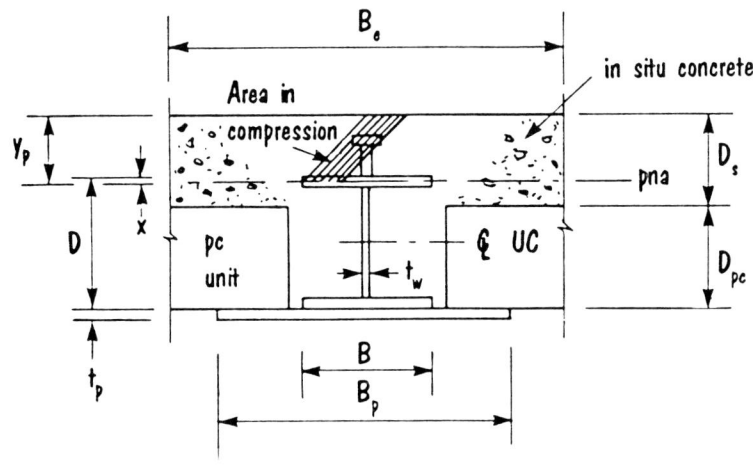

$$x = \frac{R_s + R_p - 0.45f_{cu}B_e(D_{pc} + D_s - D)}{0.45f_{cu}B_e + 2Bp_y}$$

$$y_p = D_{pc} + D_s + x - D$$

CASE 2 FS (Full Shear) - pna lies within the UC web but above UC Centre Line

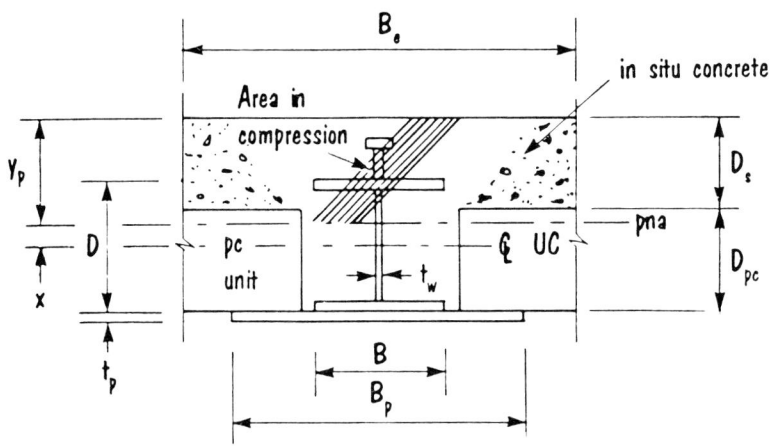

$$x = \frac{0.45f_{cu}B_e(D_s + D_{pc} - D/2) - R_p}{0.45f_{cu}B_e + 2p_y t_w}$$

$$y_p = D_s + D_{pc} - D/2 - x$$

CASE 3 FS (Full Shear) - pna lies within UC web but below UC Centre Line

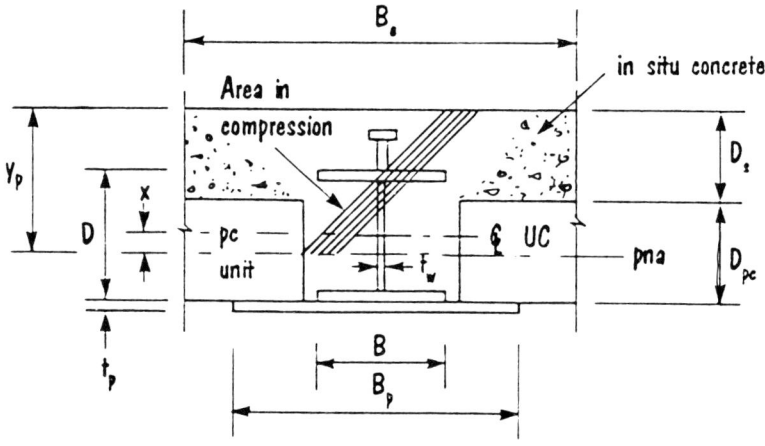

$$x = \frac{R_p - 0.45f_{cu}B_e(D_{pc} + D_s - D/2)}{0.45f_{cu}B_e + 2p_y t_w}$$

$$y_p = D_{pc} + D_s - D/2 + x$$

CASE 4 FS (Full Shear) - pna lies within bottom UC flange

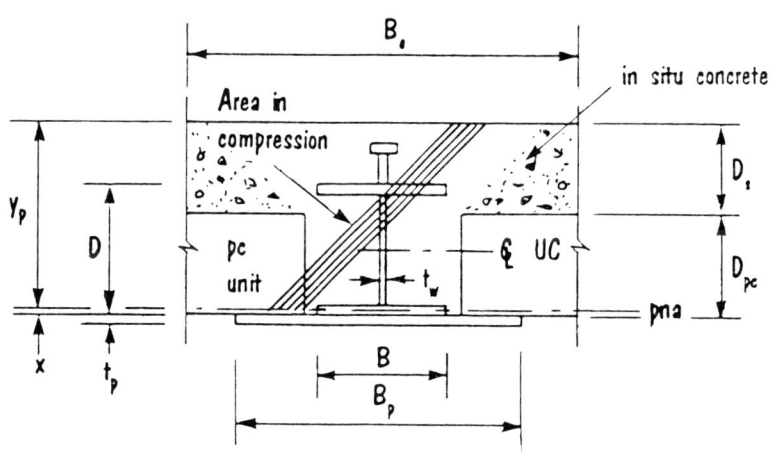

$$x = \frac{0.45f_{cu}B_e(D_s + D_{pc}) + R_s - R_p}{0.45f_{cu}B_e + 2B.p_y}$$

$$y_p = D_s + D_{pc} - x$$

CASE 1 PS (Partial Shear) : pna in top flange of UC

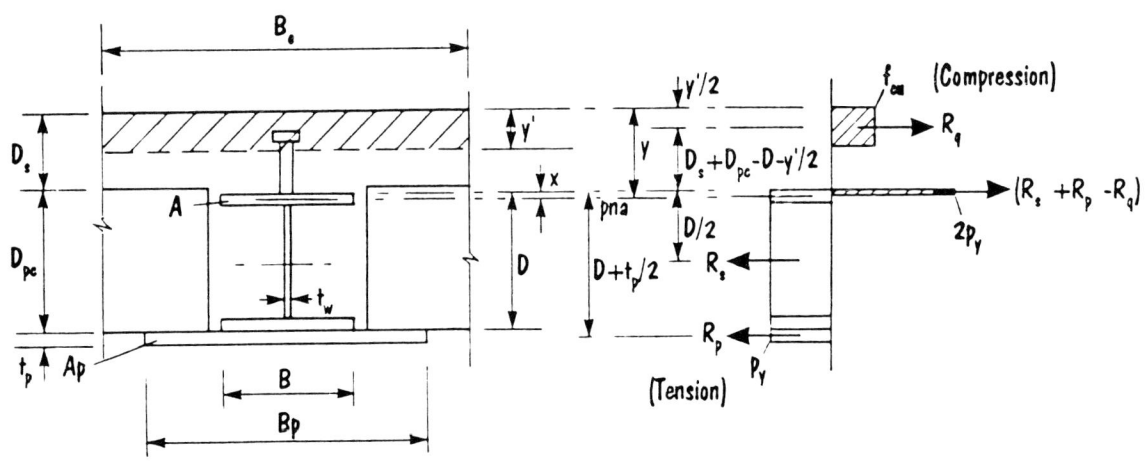

$$x = \frac{R_s + R_p - R_q}{2 B \cdot p_y}, \qquad y' = \frac{R_q}{0.45 f_{cu} B_e}$$

Moments about top flange (UC)

$$M_c = R_q (D_s + D_{pc} - D - y'/2) + \frac{R_s D}{2} + R_p (D + t_p/2) - (R_s + R_p - R_q) \frac{x}{2}$$

$$\boxed{M_c = R_q \left(D_s + D_{pc} - D \frac{R_q}{0.9 f_{cu} B_e}\right) + \frac{R_s D}{2} + R_p (D + t_p/2) - \frac{(R_s + R_p - R_q)^2}{4 \cdot B \cdot p_y}}$$

Limits for y_{crit}

If $D_{pc} \geq (D - 50)$ then $y_{crit} = D_s$
If $D_{pc} < (D - 50)$ then $y_{crit} = D_s + D_{pc} - D + 50$

See Section 4 for further information

	Job No.	PUB 810	Sheet 20 of 22	Rev.
The Steel Construction Institute	Job Title	Appendix B		
Silwood Park Ascot Berks SL5 7QN Telephone: (0344) 23345 Fax: (0344) 22944	Subject	Plastic Moment Capacity - Composite Section based on Partial Shear Connection		
CALCULATION SHEET	Client		Made by DLM	Date Oct 1991
			Checked by JWR	Date Oct 1991

CASE 2 PS (Partial Shear) : pna in web of beam, above Centre Line of UC section

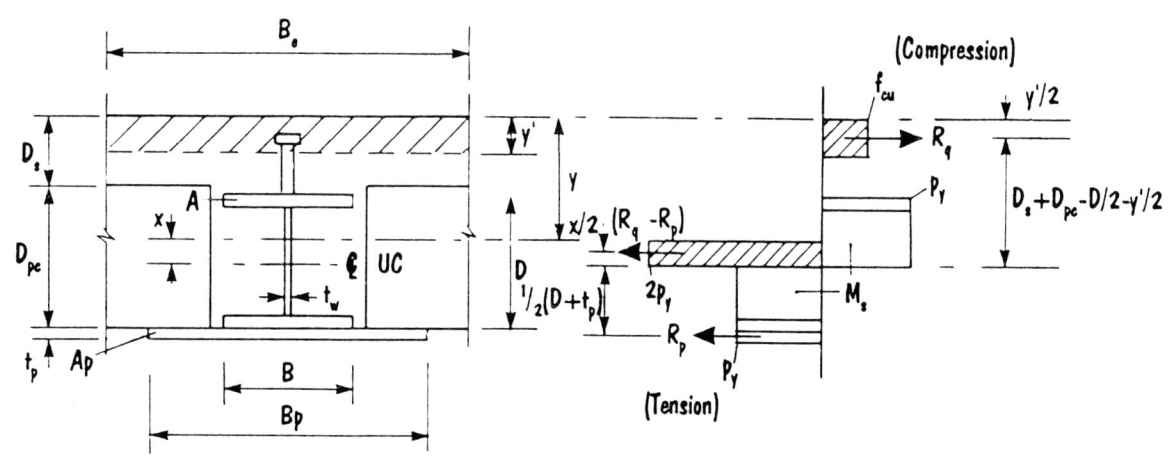

Moments about Centre Line (UC)

$$x = \frac{R_q - R_p}{2 p_y t_w}, \qquad y' = \frac{R_q}{0.45 f_{cu} B_e}$$

$$M_c = M_s + R_q (D_{pc} + D_s - D/2 - y'/2) + \frac{R_p}{2}(D + t_p) - (R_q - R_p)\frac{x}{2}$$

$$\boxed{M_c = M_s + R_q \left(D_{pc} + D_s - \frac{D}{2} - \frac{R_q}{0.9 f_{cu} B_e}\right) + \frac{R_p}{2}(D + t_p) - \frac{(R_q - R_p)^2}{4 p_y t_w}}$$

Limits for y_{crit}

If $D_{pc} \geq (D - 50)$ then $\quad y_{crit} = D_s$
If $D_{pc} < (D - 50)$ then $\quad y_{crit} = D_s + D_{pc} - D + 50$

See Section 4 for further information

CASE 3 PS (Partial Shear) : pna within the web of the UC but below the UC Centre Line

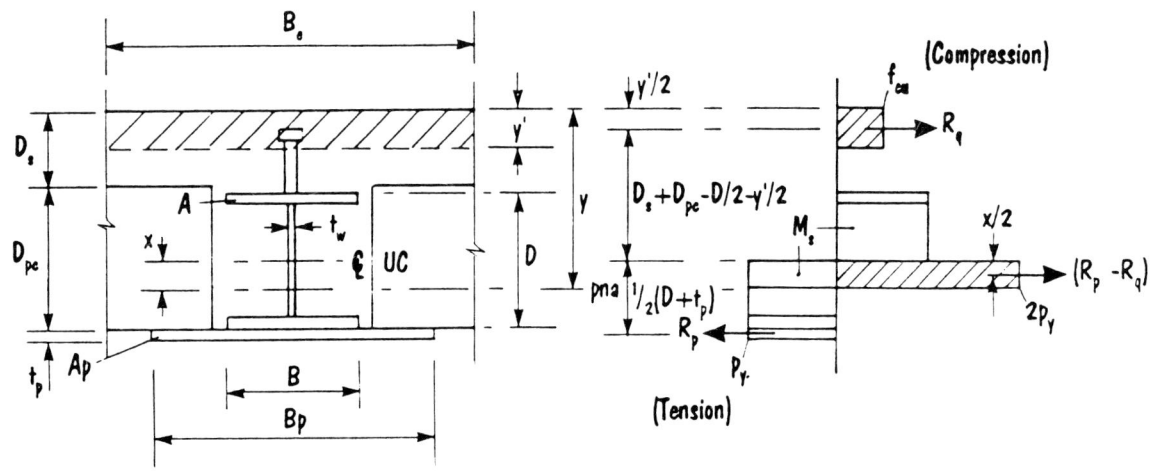

Moments about Centre Line UC

$$x = \frac{R_q - R_p}{2 p_y t_w}, \qquad y' = \frac{R_q}{0.45 f_{cu} B_e}$$

$$\boxed{M_c = M_s + R_q \left(D_{pc} + D_s - \frac{D}{2} - \frac{R_q}{0.9 f_{cu} B_e}\right) + \frac{R_p}{2}(D + t_p) - \frac{(R_p - R_q)^2}{4 p_y t_w}}$$

Limits for y'_{crit}

If $D_{pc} \geq (D - 50)$ then $\quad y'_{crit} = D_s$
If $D_{pc} < (D - 50)$ then $\quad y'_{crit} = D_s + D_{pc} - D + 50$

See Section 4 for further information

CASE 4 (Partial Shear) : pna in bottom UC flange

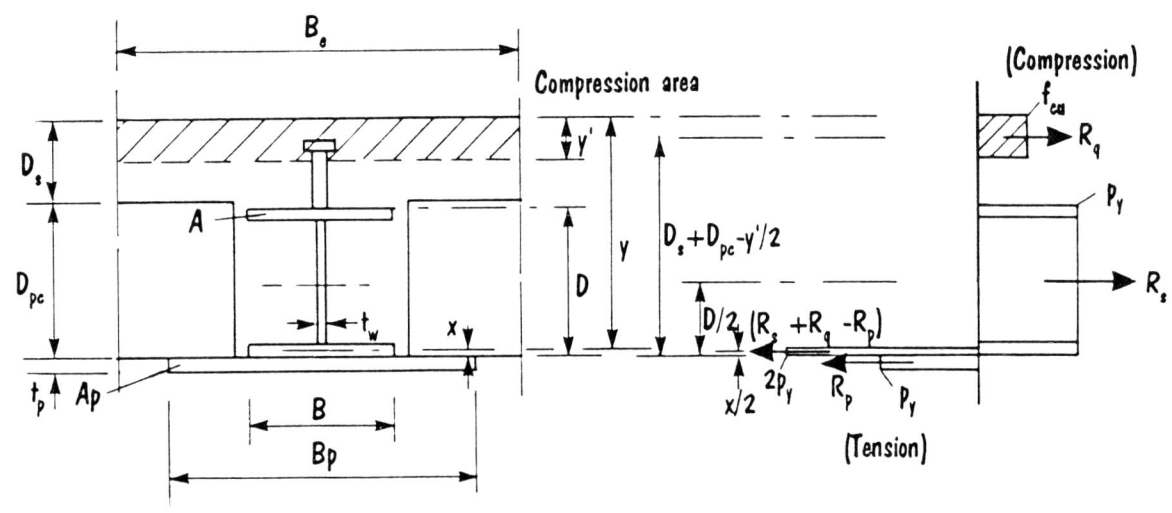

moments taken about top of bottom flange plate

$$x = \frac{R_s + R_q - R_p}{2 B p_y}, \quad y' = \frac{R_q}{0.45 f_{cu} B_e}$$

$$M_c = R_s D/2 + R_p t_p/2 + R_q (D_{pc} + D_s - y'/2) - (R_s + R_q - R_p) x/2$$

$$\boxed{M_c = \frac{1}{2} (R_s D + R_p t_p) + R_q \left(D_{pc} + D_s - \frac{R_q}{0.9 f_{cu} B_e}\right) - \frac{(R_s + R_q - R_p)^2}{4 B p_y}}$$

Limits for y'_{crit}

If $D_{pc} \geq (D - 50)$ then $\quad y'_{crit} = D_s$
If $D_{pc} < (D - 50)$ then $\quad y'_{crit} = D_s + D_{pc} - D + 50$

See Section 4 for further information

APPENDIX C: User notes for the SLIMFLOR program

Disclaimer

Although the program has been rigorously tested and produces satisfactory results, no warranty, expressed or implied, is made by the Steel Construction Institute, as to the accuracy or functioning of the program, and no responsibility is assumed in connection therewith.

The SLIMFLOR software and user notes are the property of the Steel Construction Institute. Reproduction of any kind, in whole or part in any form, without prior written consent is strictly prohibited.

Copyright: Steel Construction Institute 1991: All rights reserved.

```
------------------------------------------------------------------------
 SCI         SLIMFLOR CONSTRUCTION (INTERNAL BEAMS ONLY)    VERSION : 1.00
------------------------------------------------------------------------

          Program written by : THE STEEL CONSTRUCTION INSTITUTE
                               Silwood Park, Ascot
                               Berkshire SL5 7QN

          Distribution and Installation Support by :
                               QSE Ltd.
                               Cambrian House
                               51 Broad Street
                               Chipping Sodbury
                               Bristol, BS17 6AD
                               Tel no:  0454 319104
                               Fax no:  0454 322685

          Copyright : Steel Construction Institute 1991. All rights reserved

          THIS PROGRAM CALCULATES AND DISPLAYS THE CRITICAL DESIGN CHECKS
          PRESENTED IN THE GUIDE - SLIM FLOOR DESIGN AND CONSTRUCTION

          PRESS <C> FOR A COLOURED SCREEN OR <M> FOR A MONOCHROME SCREEN
```

An application for SLIMFLOR as a registered trademark has been made.

Contents

1. Introduction
2. Installation and running
3. Main menu
4. "Input New Data" option
5. "Save Data" option
6. "Retrieve Saved" Data option
7. "Edit Present Data" option
8. "Design Summary" option
9. "Beam Selection" option
10. "Output" option
11. Notes on Fire Input

1. Introduction

The Slim Floor Design Program (SLIMFLOR) can be run from any standard IBM compatible PC with 512KB of memory. It can be started from either a floppy disk or from a hard disk.

The monitor may be monochrome or colour.

Data entry is interactive via screen menus which can be saved for reuse. Data is checked on entry and acceptable ranges shown if values are inappropriate. The results from the design check can either be viewed on the screen or sent to a printer. However, note that sections which do not pass the design criteria can only be viewed but not printed.

2. Installation and running

2.1 Installing the program

To install the SLIMFLOR program on the hard disk or another disk, follow the instructions below:

a) Insert the SLIMFLOR disk into your floppy disk drive A: and type:

    A: \<ENTER\>

b) Start the installation by typing:

    INSTALL C: \<ENTER\>

The above process will copy the program files into a sub-directory called SLIM on the C: drive.

The program provided allows the user to install it once to a hard disk. Keep the original disk safe. It will be needed if the program is to be transferred to another machine.

2.2 Running SLIMFLOR

To run the program type:

a)   C: \<ENTER\>

b)   CD \SLIM   \<ENTER\>

c)   SLIMFLOR   \<ENTER\>

2.3 Uninstall

To uninstall the SLIMFLOR program back onto the master disk for transfer to another machine follow the following steps:

a) Insert the master disk into your floppy drive A: and type:

    A: \<ENTER\>

b)   RECALL C: \<ENTER\>

3. Main menu

The main menu page of the program consists of eight options.

Using the up and down cursor keys, the highlighted area is moved to the required option which is then selected by pressing the <ENTER> key.

Data can be either retrieved from a saved file by selecting RETRIEVE SAVED DATA option or by entering completely new data by selecting the INPUT NEW DATA option.

Once data has been retrieved or entered it can be changed or reviewed by selecting the EDIT PRESENT DATA option. The DESIGN SUMMARY option will select the first six optimum section sizes that fulfil the design criteria based on minimum weight. The BEAM SELECTION option allows all plastic or compact standard sections to be analysed.

The SAVE option saves the data to a file, whilst the OUTPUT option allows the results to be shown on screen or dumped to a printer. The END SESSION option terminates the program.

4. "INPUT NEW DATA" option

Select the INPUT NEW DATA option from the main menu. For each parameter either select the default or type in a new value followed by <ENTER>. To accept the default values press <ENTER>.

After all data has been entered, the user will be prompted to save the data to a file in a defined drive and sub-directory. Press <ENTER> to select the default drive and sub-directory.

```
----------------------------------------------------------------
SCI        SLIM FLOOR CONSTRUCTION (INTERNAL BEAMS ONLY)     VERSION : 1.0
----------------------------------------------------------------

                            MAIN MENU

                         INPUT NEW DATA

                        RETRIEVE SAVED DATA

                         EDIT PRESENT DATA

                            SAVE DATA

                          DESIGN SUMMARY

                          BEAM SELECTION

                             OUTPUT

                           END SESSION

            USE CURSOR KEYS TO MAKE SELECTION THEN PRESS <ENTER>

   JOB REFERENCE:   DEFAULT              BEAM REFERENCE:   BEAM B1
```

```
----------------------------------------------------------------
SCI        SLIM FLOOR CONSTRUCTION (INTERNAL BEAMS ONLY)     VERSION : 1.0
----------------------------------------------------------------

JOB REFERENCE (MAX 8 CHARACTERS)                       <DEFAULT > ?

BEAM REFERENCE (MAX 10 CHARACTERS)                     <BEAM B1  > ?

DESIGN TYPE (1 = Plain, 2 = Encased, 3 = Composite)         <1> ?

SPAN (1.0 - 20.0 m)                                      < 7.5> ?

SPACING (1.0 - 20.0 m)                                   < 7.5> ?
```

5. "SAVE DATA" option

Select the SAVE DATA option from the main menu or after the INPUT NEW DATA option. Press <ENTER> to select the default drive and sub-directory. This gives a list of the data files held in the chosen sub-directory. Enter either a new filename followed by <ENTER> or overwrite an existing name. There is no need to enter the . or the extension SFL. After the data is saved, the program returns to the main menu.

6. "RETRIEVE SAVED DATA" Option

Select RETRIEVE SAVED DATA option from the main menu and press R to confirm. Press <ENTER> to select the default drive and sub-directory. Type in the name of the data file to be retrieved, followed by <ENTER>. There is no need to type in the . or the extension SFL. Demo data files containing worked examples are included in the default directory. Press <ENTER> to return to the main menu.

```
--------------------------------------------------------------------------------
  SCI        SLIM FLOOR CONSTRUCTION (INTERNAL BEAMS ONLY)       VERSION : 1.0
--------------------------------------------------------------------------------
                              EXISTING DATA FILES

Z:\SLIM\M
DEMO3    .SFL      DEFAULT .SFL       DEMO1    .SFL       DEMO2    .SFL

 33132544 Bytes free




         GIVE A NAME FOR THE DATA FILE YOU WANT TO USE (MAX 8 CHARACTERS)

                  IF YOU USE AN EXISTING FILE NAME FROM THE LIST ABOVE
                      YOU WILL OVERWRITE THE PREVIOUS DATA IN THAT FILE
```

```
--------------------------------------------------------------------------------
  SCI        SLIM FLOOR CONSTRUCTION (INTERNAL BEAMS ONLY)       VERSION : 1.0
--------------------------------------------------------------------------------
                              EXISTING DATA FILES

Z:\SLIM\M
DEMO3    .SFL      DEFAULT .SFL       DEMO1    .SFL       DEMO2    .SFL

 33148928 Bytes free








                 GIVE THE NAME OF THE DATA FILE TO BE RETRIEVED
```

7. "EDIT PRESENT DATA" Option

Select EDIT PRESENT DATA from the main menu. Use the up and down cursor keys to highlight the required section of data to be changed or reviewed followed by <ENTER>.

For brevity, only the MAIN DATA edit option is described below. All other edit options are handled in a similar manner. When all changes have been made, select the RETURN TO MAIN MENU option.

Attention is drawn to Section 11 of this manual for notes on fire data and fire design.

7.1 "EDIT MAIN DATA" sub-option

To change data, use the up or down cursor keys to highlight the item and then press <ENTER> to select. Type in the new value followed by <ENTER>. Where necessary, the valid range for the parameter is shown after an inappropriate value has been entered.

If <ENTER> is pressed without any value being typed in, the previous value will be used. When all the editing is complete, highlight ACCEPT and press <ENTER>. Control is then passed back to the EDIT MENU.

```
-----------------------------------------------------------------------------
SCI        SLIM FLOOR CONSTRUCTION (INTERNAL BEAMS ONLY)       VERSION : 1.0
-----------------------------------------------------------------------------

                              EDIT MENU

                          RETURN TO MAIN MENU

                              MAIN DATA

                             CONCRETE DATA

                           SECTION DIMENSIONS

                              STEEL DATA

                             LOADING DATA

                          SHEAR CONNECTORS DATA

                              FIRE DATA

            USE CURSOR KEYS TO MAKE SELECTION THEN PRESS <ENTER>
```

```
-----------------------------------------------------------------------------
SCI        SLIM FLOOR CONSTRUCTION (INTERNAL BEAMS ONLY)       VERSION : 1.0
-----------------------------------------------------------------------------

                            EDIT MAIN DATA

          JOB REFERENCE ......................... DEFAULT

          BEAM REFERENCE ........................ BEAM B1

          CONSTUCTION DESIGN TYPE   (1)           PLAIN

          SPAN OF BEAM .......................... 7.50 m

          DISTANCE TO ADJACENT BEAM ............. 7.50 m

          BS6399 IMPOSED LOAD REDUCTION CONSIDERED (Clause 5.3)

                USE CURSOR KEYS TO SELECT THEN PRESS <ENTER>
-----------------------------------------------------------------------------

                  ACCEPT                              EDIT
```

8. "DESIGN SUMMARY" Option

Select the DESIGN option from the main menu. Select the loading condition P or F. The next screen shows the output from the "DESIGN" option. The program has performed the critical design checks detailed in the publication. The first six section sizes that pass the design criteria are shown. These have been optimised by minimum weight.

The number in the Design Criterion column shows the utilisation, whilst the letter denotes the maximum design parameter. Details of these criteria are shown below:

Design criteria

- a Construction stage: (Case 2B without slab) plastic capacity
- b Construction stage: (Case 2B with slab) plastic capacity
- c Construction stage: (Case 2B without slab) lateral torsional buckling
- d Construction stage: (Case 1A) torsion check - one sided loaded
- e Construction stage: (Case 2A) torsion check
- f Construction stage: (Case 2B with slab) ultimate combined bending check
- g In-service stage: (Case 3B) plastic capacity (+ combined bending Type 3)
- h In-service stage: (Case 3A) torsion check
- i In-service stage: (Case 3B) ultimate combined bending check
- j In-service stage: (Case 3B) ultimate shear check
- k In-service stage: dynamic sensitivity
- l In-service stage: fire resistance
- m In-service stage + construction stages: (Case 3B) combined serviceability steel tensile stresses
- n In-service stage: (Case 3B) serviceability steel compressive stress
- o Construction stage: (Case 2B with slab) serviceability steel compressive stress
- p In-service stage + construction stage: (Case 2B with slab) total serviceability steel compressive stress
- q In-service stage: (Case 3B) serviceability concrete compressive stress
- r In-service stage: (Case 3B) serviceability steel tensile stress
- s Construction stage: (Case 2B with slab) serviceability steel tensile stress
- t In-service + construction stages: (Case 3B) total serviceability steel tensile stress
- u In-service stage: (Case 3B) serviceability plate transverse stress
- v In-service stage: (Case 3B) concrete shear condition check (surface 1)
- w In-service stage: (Case 3B) concrete shear condition check (surface 2)
- * Imposed load deflection exceeds L/360

The loaded length of beam during construction depends on the loading condition P or F chosen above. If P is chosen, the maximum length of beam is shown after pre-cast (pc) units are placed on one side only. If F is chosen, the loading condition 1a is satisfied, i.e. a full internal bay can be loaded in the construction stage without placement of pc units in adjacent bays.

Press "P" to print this page, or press any other key to return to the main menu.

```
----------------------------------------------------------------------
SCI        SLIM FLOOR CONSTRUCTION (INTERNAL BEAMS ONLY)   VERSION : 1.0
----------------------------------------------------------------------



         CONSTRUCTION STAGE LOADED LENGTH (P or F)         <P> ?

       ( P = Partially loaded length,  F = Fully loaded length )


 Note :  The partially loaded condition gives the maximum length of
         beam after pc units are placed on one side only. This must
         be followed by the same length on the other side and then
         by the sequential placement of adjacent pc units.

         A fully loaded condition assumes that the units are placed
         on one side of the beam for the full span. This represents
         the most onerous condition.
```

SECTION SIZE (COMPOSITE)	STUDS NO.	DEFLECTION mm CONS	IMP	DESIGN CRITERION	CONS LOADED LENGTH, (z)	TOTAL COST
305x305x 118	42	40.64	13.77	0.99 (d)	5.04	1449
254x254x 132	48	50.28	14.90	0.99 (d)	4.88	1522
254x254x 167	46	39.81	13.11	1.00 (d)	6.72	1770
----	---	---	---	---	---	---
----	---	---	---	---	---	---
----	---	---	---	---	---	---

d Construction stage : (Case 1A) torsion check - one side loaded

Refer to Appendix D in Design Guide for further explanation of above criteria

******** PRESS <P> TO PRINT OR ANY OTHER KEY TO RETURN TO MAIN MENU *********

9. "BEAM SELECTION" option

Select the BEAM SELECTION option from the main menu. Select the loading condition P or F.

Use the up and down cursor keys to select a section to be checked against the design rules. The sections shown with an asterisk are those that were optimised in the DESIGN option, press <ENTER> to confirm the selection.

The section is then analysed and either passes or fails. There are four failure modes:

(a) Semi-compact section. This section is not recommended and no further analysis is performed.

(b) Section too small. The section size chosen should be increased. Output from the analysis is available on screen only.

(c) Section failed. This section failed to comply with one or more of the design criteria. All the design criteria are analysed and output is available on screen only.

(d) Degree of shear connection less than 40% due to inadequate concrete depth. This section is not recommended and no further analysis is performed.

Full output for passed sections is available both on screen and on printer.

Press <ENTER> to return to the main menu.

```
---------------------------------------------------------------------------
  SCI       SLIM FLOOR CONSTRUCTION (INTERNAL BEAMS ONLY)      VERSION : 1.0
---------------------------------------------------------------------------


    SECTIONS FROM THE DESIGN SUMMARY ( marked with * ) :

                              *305x305x 118
                              *254x254x 132
                              *254x254x 167




         CURRENT SELECTION :      *305x305x 118

         USE VERTICAL CURSOR KEYS TO MAKE SELECTION THEN PRESS <ENTER>  <<
```

10. "OUTPUT" option

This option will display the critical calculations shown in the publication for the chosen section. Highlight this option and press <ENTER> to confirm.

Press S to display the results on the screen or press P to send the output to a printer. Note, only those sections that passed will be available for output to a printer. Failed section output is available on the screen only.

After displaying or printing the output, the program returns to the main menu.

If the input data has changed remember to save the data by selecting the "SAVE DATA" option. Exit the program by selecting the "END SESSION " option.

11. Fire design

The part of the program dealing with fire is reasonably self explanatory but certain points may require clarification.

(a) Fire resistance period.
   Although the user may enter 30, 60 or 90 minutes a moment capacity check is only carried out for 60 minutes.

(b) Plate thickness
   A moment capacity check is only carried out for plates at least 15 mm thick.

(c) Imposed load
   In fire, imposed loads may have a partial load factor of 0.8 or 1.0. The user must enter the amount of load at the factor of 0.8. The remainder of the imposed load is assumed to have a partial load factor of 1.0.

   If the user changes the imposed load for normal design, the non-permanent imposed load for fire will be automatically set to zero and must therefore be re-entered if required. A warning is displayed.

(d) Output
   For 60 minutes fire resistance, the moment capacity and applied moments are printed together with the load ratios required and achieved. If no moment capacity check is carried out, one of the following messages will appear:

   - NO CHECKS NECESSARY FOR 30 MINUTES
   - PLATE < 15 mm - NO FIRE CHECK AVAILABLE (60 and 90 minutes only)
   - PLATE SHOULD BE FIRE PROTECTED FOR 90 MINUTES

(e) Design and failure criteria
   The moment capacity in fire will only be considered as a design or failure criteria for 60 minutes with plates not less than 15 mm thick.

```
-------------------------------------------------------------------------------
SLIMFLOR CONSTRUCTION (INTERNAL BEAMS ONLY)                    VERSION : 1.00
-------------------------------------------------------------------------------
JOB REFERENCE    Demo3          FILE REFERENCE demo3.SFL       11/11/1991
-------------------------------------------------------------------------------
BEAM REFERENCE   Type 3                                        PAGE   1 OF 11
-------------------------------------------------------------------------------

   FLOOR PLAN DATA :
   -----------------
   Span                         8.00 m      Adjacent beams         8.00 m
-------------------------------------------------------------------------------
   CONCRETE DATA : (Calculation based on overall construction depth)
   ---------------
   Cube strength                30 N/mm2    Concrete depth         145 mm
   Wet density                  24 kN/m3    Modular ratio          10
-------------------------------------------------------------------------------
   SECTION DIMENSIONS :
   --------------------
   Section size   *305x305x 118
   Beam depth                   314.5 mm    Web thickness          11.9 mm
   Flange width                 306.8 mm    Flange thickness       18.7 mm
   Plate width                  506.8 mm    Plate thickness        15.0 mm
   Distance  a                   62.5 mm    Distance  b            37.5 mm
   Depth of PC units            260.0 mm    DESIGN TYPE    (3)   COMPOSITE
-------------------------------------------------------------------------------
   STEEL DATA :
   ------------
   Steel grade                     50       Beam design strength   345 N/mm2
   Modulus of elasticity       205000 N/mm2 Plate design strength  355 N/mm2
   Cost of fabricated steel       850 /Tonne Cost of one stud       1.0
-------------------------------------------------------------------------------
   LOADING DATA :
   --------------
   Imposed load : (occupancy)   5.00 kN/m2  (partitions)           1.00 kN/m2
   Total imposed load with BS6399 reduction                        5.62 kN/m2
   Services and finishes        0.40 kN/m2  Weight of PC units     4.00 kN/m2
   Dead load factor              1.4        Imposed load factor    1.6
-------------------------------------------------------------------------------
   SHEAR CONNECTORS DATA :
   -----------------------
   Diameter of studs             19 mm      Height of studs         70 mm
   Design capacity               70 kN
-------------------------------------------------------------------------------
```

COMPUTER PROGRAM "SLIMFLOR"

An interactive computer program, suitable for any IBM compatible PC, has been developed by SCI to perform the recommended calculations in this design guide.

See Reverse  For Ordering Details

The software is available in either 5.25" or 3.5" format.

Full price is £120.00 plus VAT
SCI Corporate Member price is £80.00 plus VAT
Discounts are available for multiple copies.
Price includes delivery

See Appendix C for the user manual.

> **Please complete this form and post with cheque to:-**
> QSE Ltd, Cambrian House, 51 Broad Street,
> Chipping Sodbury, Bristol, BS17 6AD.
> Please supply ( ) copy of SLIMFLOR on
>
> ☐ 3.5" format
>
> ☐ 5.25" format
>
> Full price : £141.00 inc. VAT.
> SCI Corporate Member price £94.00 inc. VAT.
>
> I enclose a cheque made payable to QSE for £............
>
> Name............................... Position........................
> Company...
> Address..
>   ....
>   ....
>   ....
> Tel............................ Fax............................
> Date...........................
> SCI Corporate Membership Number........................
> For technical enquiries regarding the software, please contact the SCI.

Terms of Condition

Although the program has been rigorously tested and produces satisfactory results, no warranty expressed or implied, is made by The Steel Construction Institute, as to the accuracy or functioning of the program, and no responsibility is assumed in connection therewith.

The SLIMFLOR software and manual are the property of The Steel Construction Institute.

Copyright: The Steel Construction Institute 1991. All rights reserved.

NOTES

NOTES

NOTES

Typeset and page make-up by The Steel Construction Institute, Ascot, Berks. 1000/12/91
Presswork and binding by Hollen Street Press Limited, Slough, Berks.